SpringerBriefs in Computer Science

Series Editors

Stan Zdonik
Peng Ning
Shashi Shekhar
Jonathan Katz
XindongWu
Lakhmi C. Jain
David Padua
Xuemin Shen
Borko Furht
V.S. Subrahmanian
Martial Hebert
Katsushi Ikeuchi
Bruno Siciliano

For further volumes:
http://www.springer.com/series/10028

Jin Tang • Yu Cheng

Intrusion Detection for IP-Based Multimedia Communications over Wireless Networks

Jin Tang
AT&T Labs
Warrenville
IL, USA

Yu Cheng
Department of Electrical and Computer
Engineering
Illinois Institute of Technology
Chicago, IL, USA

ISSN 2191-5768   ISSN 2191-5776 (electronic)
ISBN 978-1-4614-8995-5   ISBN 978-1-4614-8996-2 (eBook)
DOI 10.1007/978-1-4614-8996-2
Springer New York Heidelberg Dordrecht London

Library of Congress Control Number: 2013949152

Printed on acid-free paper

Springer is part of Springer Science+Business Media (www.springer.com)

To my wife Huan—Jin
To my wife Yanning and our daughter Annabelle—Yu

Preface

IP-based multimedia communications have become prevailing in recent years. At the same time, with the increasing coverage of the IEEE 802.11^{TM}-based wireless networks, IP-based multimedia communications over wireless networks are drawing extensive attention in both academia and industry. However, due to the openness and distributed nature of the protocols involved, such as the session initiation protocol (SIP) and the IEEE 802.11^{TM} standard, it becomes easy for malicious users in the network to achieve their own gain or disrupt the service by deviating from the normal protocol behaviors. This book presents real-time intrusion detection techniques that can quickly track down the malicious behaviors which manipulate the vulnerabilities from either the 802.11^{TM} or the SIP protocols.

Specifically, for the intrusion detection over the 802.11^{TM} protocol, a real-time detector exploiting the nonparametric cumulative sum (CUSUM) test is designed to quickly find a selfish malicious node without any a priori knowledge of the statistics of the selfish misbehavior. While most of the existing schemes for selfish misbehavior detection depend on heuristic parameter configuration and experimental performance evaluation, this book presents a Markov chain-based analytical model to systematically study the CUSUM-based detector, for guaranteed performance in terms of average false positive rate, average detection delay, and missed detection ratio. Further, to achieve better detection performance, by enhancing the FS detector, an adaptive detector is developed with the Markov decision process (MDP). Then based on a reward function defined in this book, an optimal decision policy can be determined to maximize the overall system benefit through a linear programming formulation. The optimal policy also indicates the operation of the adaptive detector, which yields better performance in both false positive rate and detection delay.

For attacks on the SIP layer, this book first focuses on the well-known flooding attack and develops an online scheme to detect and subsequently prevent the attack, by integrating a novel three-dimensional sketch design with the Hellinger distance detection technique. A very challenging attack, the stealthy attack, is also addressed in this book. In a stealthy attack, intelligent attackers can afford a long time to attack the system and only incur minor changes to the system within each sampling period.

A wavelet-based technique is presented to effectively deal with the stealthy attack. Moreover, a new type of malformed message attack, which manipulates both the "Session-Expires" header in the SIP message and openness of wireless protocols to severely drain the network resources, is also addressed.

In summary, this book presents interdisciplinary techniques to achieve an effective real-time intrusion detection system, which interleaves medium access control (MAC) protocol analysis, CUSUM-based detector design, a novel Markovian model for CUSUM detectors, Markov decision process-based performance optimization, sketch-based traffic modeling, and wavelet-based signal processing techniques.

Chicago, IL, USA   Jin Tang and Yu Cheng

Contents

Chapter 1
Introduction

IP-based multimedia communications have become prevailing in recent years, with a wide range of benefits such as cost-efficient deployment, plenty of features, and convenience for service integration. At the same time, the deployment of IEEE 802.11^{TM} based wireless networks has been dramatically increasing over years due to their high-speed access, easy-to-use feature, and economical advantages. The convergence of such two trends, IP-based multimedia communications over 802.11^{TM} based wireless networks, leads to a promising all-IP platform to provision economic high-quality multimedia services to mobile users anytime and anywhere, which has drawn extensive attentions from both academia and industry.

The negative side is that IP-based multimedia communications, including one of its most prominent applications, voice over IP (VoIP), are technologies based on open standards like the session initiation protocol (SIP) [39]; thus they are likely to have more security concerns than the traditional public switched telephone networks (PSTN) [6, 9, 11, 13, 16–18, 24, 31, 32, 40, 42, 43, 45, 52, 53]. Attacks like SIP flooding [42, 44, 52] and malformed messages [15, 52] are easy to launch and capable of quickly draining the resource of both network nodes and mobile users. The attacks disrupt perceived quality of service (QoS) and subsequently lead to denial of service (DoS). When multimedia communications over wireless is considered, the IEEE 802.11^{TM} [56] (as one of the most popular wireless network standards) bears its own problems [2, 4, 8, 19, 21, 22, 25–27, 29, 33, 34, 36, 50, 51]. As a contention-based protocol, it assumes that every participant in the network acts in compliance with the protocol rules to ensure that each node gets a reasonably fair share of the network [3, 10, 14, 55]. In particular, every node should access the network in a cooperative manner and randomly delays transmissions to avoid collisions by following a common backoff rule. However, in such a distributed environment without a centralized controller, a malicious node may deliberately choose a smaller backoff timer and selfishly gain an unfair share of the network throughput at the expenses of other normal nodes' channel access opportunities [5, 7, 35, 37, 48, 49]. Moreover, only to make things worse, the wireless network devices nowadays have developed to be easily programmable and reconfigurable [47, 49], which lets malicious actions become much more feasible and not rely on much expert knowledge.

J. Tang and Y. Cheng, *Intrusion Detection for IP-Based Multimedia*
Communications over Wireless Networks, SpringerBriefs in Computer Science,
DOI 10.1007/978-1-4614-8996-2_1, © The Author(s) 2013

As a result of the aforementioned problems, IP-based multimedia communications over an IEEE 802.11™ based wireless network can be deteriorated easily from various aspects. Thus in this book, we aim to develop real-time intrusion detection techniques without modification to the relevant standard protocols, which can quickly track down the malicious behaviors manipulating the vulnerabilities from either multimedia communications or 802.11™ protocols.

Generally, intrusion detection systems are classified into two major approaches, *signature based* and *behavior based*. The signature-based approach profiles known network attacks as signatures. Detection systems in this approach raise alert if the on-going traffic patterns match the profiled signatures. It can accurately identify known attacks but is not able to detect new anomalies. Rather than profiling known attacks, the behavior-based approach builds models that represent normal behaviors on the network. Alarms are raised if the observed behaviors significantly deviate from the behaviors estimated by the model. The main advantages of the behavior-based approach are that a priori knowledge of attack strategies is not required and new anomalies unknown before can be detected. Our detection methodology in this book will adopt the behavior-based approach.

1.1 Selfish Misbehavior Detection in 802.11™

1.1.1 FS Detector and Markov Chain Based Analytical Model

To efficiently detect the selfish misbehavior in the IEEE 802.11™ based wireless networks, a detection scheme needs to address the two main correlated challenges: (1) *unknown misbehavior strategy*, (2) *real-time detection of the misbehavior*. For the first challenge, since a malicious node can first behave as a normal node and then manipulate its backoff timer to a random small value at any time, we have no way to know the misbehavior strategy a priori. For the second, the misbehavior needs to be detected in real time and we can then isolate the malicious node to prevent it from bringing more harm to the network as soon as possible. The existing solutions either can not address both issues at the same time [35, 36, 38, 48], or require modifications to the 802.11™ protocols [29, 30].

Addressing the challenges, in this book we first design a real-time backoff misbehavior detector, termed as the *fair share detector* (FS detector), which exploits the non-parametric cumulative sum (CUSUM) test to quickly find a selfish malicious node without any a priori knowledge of the statistics of the selfish misbehavior. The FS detector takes each successful transmission over the network as its observation sample. Such a sampling method is independent of the network size and turns out to result in good performance in both false positive rate and detection delay. Also, the FS detector does not require any modification to the protocols, and can be implemented by any node assuming the role of the detection agent that monitors the network.

Another significant open research issue regarding the selfish misbehavior detection is that most of the existing detection schemes depend on heuristic parameter configuration and experimental performance evaluation. Such a heuristic approach largely limits the flexibility and robustness of the detection scheme; a change of the operation context could trigger the retraining of the configuration parameters by experimenting over a large set of data traces and the performance under those heuristic parameters is not theoretically provable.

To address the issue, in this book we further develop an analytical model for the FS detector, which can provide quantitative performance analysis and theoretical guidance on system parameter configuration. Specifically, we use a discrete-time Markov chain to model the behavior of the detector, because the detector's next state depends only on its current value and the coming observation sample. This Markov chain based model enables us to conduct rigorous quantitative analysis of the FS detector on three fundamental metrics: *average false positive rate*, *average detection delay*, and *missed detection ratio*, and further compute the system configuration for guaranteed performance. In particular, the Markov chain modeling the FS detector takes different transition probabilities under the normal traffic condition and under the abnormal condition with misbehaving nodes present, respectively. The Markov chain obtained from the normal traffic condition can be used to directly calculate the average false positive rate and also provide the initial states for misbehavior analysis. Based on these initial states, we can then use the Markov chain under the abnormal conditions to analyze the average detection delay and the missed detection ratio. Note that the missed detection ratio is not often considered in the context of CUSUM test due to its "non-stop until detection" property. In this book, we examine a *missed detection ratio under a detection delay constraint*, which is of importance regarding real-time detection.

1.1.2 Adaptive Detector and Markov Decision Process Based Modeling

An adaptive detector is necessary in practice, as it needs to deal with different misbehaving scenarios where the number of selfish nodes and the contention windows exploited by each selfish node are different. In particular, based on the operation of the FS detector, upon an observation sample, the adaptive detector further needs to determine an action of whether to more aggressively increase or decrease its value. At a state, if the observed node is inferred to be a misbehaving one with a high probability, aggressively increasing the detector value can lead to a smaller detection delay while not impacting the false positive rate much. On the other hand, if the observed node is inferred to be a normal one with a high probability, aggressively decreasing the detector value can mitigate the false positive rate without impacting the detection delay much.

Regarding the adaptive detector, a proper decision on the action at each state is crucial for improving the detection performance, as bad decisions may degrade the system performance. We have indicated that our CUSUM-based FS detector can be modeled as a discrete-time Markov chain. Thus, we can further resort to the Markov decision process (MDP) technique to guide the adaptive design. In an MDP formulation, the optimal action for each detection state will be obtained by solving an optimization problem to maximize the benefit based on a reward function. This book develops a novel reward function which will generate a positive reward for a correct decision and a negative penalty for a wrong decision (for example, aggressively increasing the detector value when monitoring a normal node). The Markovian model also enables us to theoretically analyze the detection performance in terms of average false positive rate, average detection delay, and missed detection ratio. Efficiency of the MDP-based adaptive design is demonstrated with comparison to the performance of the FS detector in which no adaptive action is taken in each state. To the best of our knowledge, we are the first applying the MDP to improve the performance of the CUSUM-based detector.

1.2 SIP Layer Attack Detection

1.2.1 Flooding Attack Detection

A natural idea for flooding detection is to identify changes in traffic volume/rate since an unreasonable volume/rate burst can well imply some malicious behavior on the network [28, 41]. However, one major limitation of volume/rate-based monitoring is that low-rate flooding can hardly be distinguished from the normal rate fluctuation due to randomness. Fortunately, besides just minor volume/rate changes, anomalies are likely to induce different probability distributions from the normal one, which reveals the presence of anomalies. The Hellinger distance (HD) [54] is a well-known metric to describe the deviation between two probability distributions, which has been used in [44] to implement a flooding detection system with good sensitivity. However, the scheme in [44] establishes a probability distribution by monitoring the relative proportions of four types of SIP messages associated with four SIP attributes within the total traffic. The detection method will become ineffective if the four attributes are proportionally flooded simultaneously. We refer to such an attack as *multi-attribute attack*. Also through investigation, we find that as there is a relatively large time difference between the BYE attribute and the other three attributes due to call holding times, dynamic normal traffic arrivals can severely undermine the effectiveness of the scheme in [44]. Moreover, the scheme does not address the important issues of how to protect the detection threshold from being polluted by attacks and how to subsequently prevent the attacks after detection.

In this book, we develop a versatile scheme for detecting and preventing the SIP flooding attacks in IP-based multimedia communications networks, by integrating the sketch technique [20, 28] with the HD-based detection for a more effective and flexible solution. Sketch is capable of summarizing each of the incoming

SIP messages into a compact and fixed-size data set by random hash operations. Based on the sketch data set, we can establish a probability distribution for each SIP attribute independently, termed as *sketch data distribution*, which is the cornerstone of our design. Especially, we design a generic *three-dimensional sketch*: the sketch comprises multiple two-dimensional *attribute hash-tables* (one for each SIP attribute), and each attribute table consists of multiple *element hash-rows* (one associated with a different hash function). The three-dimensional sketch design allows us to apply HD detection to examine the anomaly over each SIP attribute separately and therefore successfully resolve the multi-attribute attack. The multiple element hash-rows provision a voting scheme to improve detection accuracy. Also due to the separate examination on each attribute, the time difference between the attributes does not affect our scheme and we are able to maintain high detection accuracy under dynamic normal traffic arrivals. Furthermore, the multiple hash-row design with an attribute table can be leveraged to identify the offending SIP messages responsible for the flooding attack over the attribute under consideration. We can then selectively discard those messages to efficiently prevent the attack. In addition, we develop an *estimation freeze mechanism* that can protect the HD threshold estimation from being impacted by the attacks. A side benefit of the estimation freeze mechanism is that the durations of attacks can be identified.

1.2.2 Stealthy Attack Detection

One special form of the SIP flooding attacks is called the stealthy attack, where intelligent attackers deliberately increase the flooding rates in a slow pace. Although slow and seemingly unnoticeable, the attack is still able to gradually degrade the processing capability of the targets and bring serious damages to the network. As the attacking rate only increases slightly or even keeps the same in consecutive sampling periods, the Hellinger distance based detection schemes are not able to effectively identify changes in traffic.

Realizing this limitation, in this book, we continue to propose a scheme based on the wavelet analysis to address quick detection of the stealthy SIP flooding attack. The fundamental reason for the ineffectiveness of the HD-based detection schemes is that they fail to identify the deviation from normal traffic brought by the attack. Wavelet is a signal processing technique which can extract information from the raw traffic signal by decomposing the signal to enable observations on different levels, i.e., approximation signal and detail signal. This is also known as multi-resolution analysis (MRA). In the proposed scheme, we monitor the percentage of energy corresponding to the detail signal obtained from the wavelet analysis, which keeps low under the normal traffic condition but will increase sharply after the attack starts. This enables the scheme to expose the deviation brought by the stealthy attack in real time and achieve quick detection even though the attack only slowly influences the network traffic. Moreover, considering the scalability of the proposed scheme, we again resort to the sketch technique, which can summarize the traffic observations to a fixed-size hash table regardless of how many users exist in the network. Sketch provides raw traffic signals for the wavelet analysis and has crucial impact on attack detection.

1.2.3 Resource-Drained Malformed Message Attack Detection

SIP defines open-ended control message implementation, users are able to modify the values of some SIP headers as they like. By realizing this and also the vulnerability of wireless networks, we identify a resource-drained malformed message attack. The basic SIP specifications do not require the proxy servers to track states for established sessions, thus a basic proxy server is not able to determine whether a session is alive or dead after it is established. As a result, the server will even hold resources reserved for failed sessions. To address this problem, an extension header field "Session-Expires", namely the session timer [12] has been proposed for SIP to serve as a keep-alive mechanism. It allows the proxy server to keep resources for established sessions just as long as the specified session timers if the sessions are not explicitly terminated by BYE messages. In normal situation, the "Session-Expires" header is used by SIP user agents to inform the proxy server the session durations. The proxy server will then assign resource for the session according to the specified amount of time. However, by utilizing this fact, an attacker can set an arbitrarily large value for the header and keep holding resource in the proxy server until the session timer expires. Also, because it is in wireless networks, the attacker can easily disconnect from the network and still hold resources on the proxy. By simply repeating the actions above, an attacker is able to gradually drain the resources on a proxy server and cause denial of service (DoS) to other normal users. The attack requests are sent to the proxy server following regular traffic rates, thus they can hardly be detected by existing volume-based intrusion detection systems. More severely, the attacks can cause even greater harm if they are initiated from distributed sources on a wireless network, i.e., the DDoS attack. In this case, resources of the SIP proxy servers will be drained at a much faster pace.

As a counter-measure, we propose a detection scheme for the resource-drained attack based on the statistical Anderson–Darling test [1, 46] through investigating the characteristics of both the normal and attack behaviors. The "Session-Expires" header conveys the duration of the session and there is no default value defined for the header [12]. Thus the header value can be modeled as a random variable which has the same distribution of the session holding times. According to existing research, VoIP, as one of the most prominent IP-based multimedia communications applications, has its session holding times closely following a log-normal distribution to reflect the long tail characteristic [23]. Therefore the distribution of the "Session-Expires" header values can also be modeled using a log-normal distribution in the normal situation. Attackers set arbitrarily long values for the session timers to maximize the attack effect. Although unpredictable, their behavior is still very likely to introduce deviation from the log-normal distribution of the session timers in normal situation. From this insight, we build our detection scheme based on one of the most powerful tools in normality test, the Anderson–Darling test. We monitor the session timer values from the whole network as the input of the scheme. Specified parameter values of the session timer distribution are not required due to the strength of the Anderson–Darling test. The scheme is robust against any attacks deviating from the normal behavior. And the DDoS attacks are even easier to be detected as they will induce more significant deviation from the normal session timer distribution.

1.3 Overview of This Book

There are four chapters beyond this chapter. Chapters 2 and 3 address detection of the selfish misbehavior in 802.11^{TM}. In Chap. 2, we propose the FS detector based on the non-parametric CUSUM test. Further, a Markov chain based analytical model is developed to systematically study the FS detector, which enables us to quantitatively compute the system configuration parameters for guaranteed performance. Simulation results are also presented to confirm the accuracy of our theoretical analysis. In Chap. 3, we enhance the FS detector to develop an adaptive detector with the Markov decision process. Both theoretical analysis and simulation results are provided to demonstrate the performance of the adaptive detector. Chapters 4 and 5 investigate the detection of various SIP layer attacks. Chapter 4 addresses the flooding attack detection by integrating a novel three-dimensional sketch design with the Hellinger distance detection technique. Chapter 5 focuses on the stealthy attack detection through the wavelet analysis, and the detection of the resource-drained malformed message attack identified by us, utilizing the Anderson–Darling test. The effectiveness of the corresponding detection schemes are demonstrated through simulation results. For readers' convenience, we also include a related work section at the end of Chaps. 3 and 5 as further reading on the literature of selfish misbehavior detection in 802.11^{TM} and SIP layer attack detection, respectively.

References

1. T. Anderson and D. Darling, "Asymptotic Theory of Certain "Goodness-of-Fit" Criteria Based on Stochastic Processes," *Annals of Mathematical Statistics*, 1952.
2. J. Bellardo, and S. Savage, "802.11 Denial-of-Service Attacks: Real Vulnerabilities and Practical Solutions," in *Proc. USENIX Security Symposium*, 2003.
3. G. Bianchi, "Performance Analysis of the IEEE 802.11 Distributed Coordination Function," *IEEE Journal on Selected Areas of Communication*, vol. 18, no. 3, pp. 535–547, Mar. 2000.
4. L. Buttyan and J. Hubaux, "Report on a Working Session on Security in Wireless Ad hoc Networks," in *Mobile Computing and Communications Review*, vol. 6, no. 4, 2002.
5. M. Cagalj, S. Ganeriwal, I. Aad and J. Hubaux, "On Selfish Behavior in CSMA/CA Networks," in *Proc. IEEE INFOCOM*, 2005.

6. C. Callegari, R. Garroppo, S. Giordano, M. Pegano and F. russo "A Novel Method for Detecting Attacks Towards the SIP Protocol," in *Proc. IEEE International Symposium on Performance Evaluation of Computer and Telecommunication Systems*, 2009, pp. 268–273.

7. A. Cardenas, S. Radosavac and J. Baras, "Detection and Prevention of MAC Layer Misbehavior in Ad Hoc Networks," in *Pro. ACM 2nd Workshop on Security of ad hoc and Sensor Networks*, 2004, pp. 17–22.

8. A. Cardenas, S. Radosavac and J. Baras, "Evaluation of Detection Algorithm for MAC Layer Misbehavior: Theory and Evaluation," *IEEE Trans. Networking*, vol. 17, no. 2, pp. 605–617, Apr. 2009.

9. E. Chen, "Detecting DoS Attacks on SIP Systems," in *Proc. 1st IEEE Workshop on VoIP Management and Security*, 2006, pp. 53–58.

10. Y. Cheng, X. Ling, W. Song, L. Cai, W. Zhuang, and X. Shen, "A Cross-layer Approach for WLAN Voice Capacity Planning," *IEEE Journal on Selected Areas of Communications*, vol. 25, no. 4, pp. 678–688, May 2007.

11. Y. Ding and G. Su, "Intrusion Detection System for Signal Based SIP Attacks Through Timed HPCN," in *Proc. 2nd International Conference on Availability, Reliability and Security*, 2007, pp. 190–197.

12. S. Donovan, and J. Rosenberg, "Session Timers in the Session Initiation Protocol (SIP)," IETF RFC 4028, Apr. 2005.

13. S. Elhert, C. Wang, T. Magedanz and D. Sisalem, "Specification-Based Denial-of-Service Detection for SIP Voice-over-IP Networks," in *Proc. 3rd IEEE International Conference on Internet Monitoring and Protection*, 2008, pp. 59–66.

14. C. Foh and M. Zukerman, "Performance Analysis of the IEEE 802.11 MAC Protocol," in *Proc. Europe Wireless*, 2002, pp. 184–190.

15. D. Geneiatakis, G. Kambourakis, T. Dagiuklas, C. Lambrinoudakis and S. Gritzalis, "A Framework for Detecting Malformed Messages in SIP Networks," in *Proc. 14th IEEE Workshop on Local and Metropolitan Area Networks*, 2005.

16. D. Geneiatakis, G. Kambourakis, T. Dagiuklas, C. Lambrinoudakis and S. Gritzalis, "SIP Security Mechanism: A State-of-the-Art Review," in *Proc. 5th International Network Conference*, 2005, pp. 147–155.

17. D. Geneiatakis, G. Kambourakis, C. Lambrinoudakis, T. Dagiuklas and S. Gritzalis, "SIP Message Tampering: THE SQL code INJECTION attack," in *Proc. IEEE 13th International Conference on Software, Telecommunications and Computer*, 2005.

18. D. Geneiatakis, T. Dagiuklas, G. Kambourakis, C. Lambrinoudakis, S. Gritzalis, K. S. Ehlert and D. Sisalem, "Survey of Security Vulnerabiliteis in Session Initiation Protocol," *IEEE Communication Surveys & Tutorials*, vol. 8, no. 3, pp. 68–81, 2006.

19. L. Giarre, G. Neglia and I. Tinnirello "Medium Access in WiFi Networks: Strategies of Selfish Nodes," in *IEEE Signal Processing Magazine*, pp. 124–128, Sept. 2006.

20. A. Gilbert, S. Guha, P. Indyk, S. Muthukrishnan and M. Strauss, "Quicksand: Quick Summary and Analysis of Network Data," DIMACS Technical Report 2001–43, 2001.

21. L. Guang, C. Assi and A. Benslimane, "Enhancing IEEE 802.11 Random Backoff in Selfish Environment," in *IEEE Trans. Vehicular Technology*, vol. 57, no. 3, pp. 1806–1822, May 2008.

22. V. Gupta, S. Krishnamurthy and M. Faloutsos, "Denial of Service Attacks at the MAC Layer in Wireless Ad Hoc Network," in *Proc. IEEE MILCOM*, 2002.

23. F. Gustafson and M. Lindahl, "Evaluation of statistical distributions for VoIP traffic modelling," University Essay from University West, Department of Economics and IT, 2009.

24. Y. He, Y. Wen and H. Zhao "SPIT Detection and Prevention Method in VoIP Environment," in *Proc. IEEE 3rd International Conference on Availability, Reliability and Security*, 2008, pp. 473–478.

25. M. Heusse, F. Rousseau, G. Berger-Sabbatel and A. Duda, "Performance Anomaly of 802.11b," in *Proc. IEEE INFOCOM*, 2003.

26. Y. Jin and G. Kesidis "Distributed Contention Window Control for Selfish Users in IEEE 802.11 Wireless LANs," in *EEE Journal on Selected Areas in Communication*, vol. 25, no. 6, pp. 1113–1123, Aug. 2007.

27. J. Konorski, "Solvability of a Markovian Model of an IEEE 802.11 LAN under a Backoff Attack," in *Proc. 13th IEEE International Symposium on Modeling, Analysis, and Simulation of Computer and Telecommunication Systems*, 2005, pp. 491–498.

28. B. Krishnamurthy, S. Sen, Y. Zhang and Y. Chen, "Sketch-based Change Detection: Methods, Evaluation, and Applications," in *Proc. ACM SIGCOMM IMS*, 2003.

29. P. Kyasanur and N. Vaidya, "Detection and Handling of MAC Layer Misbehavior in Wireless Networks," in *Proc. IEEE DSN*, 2003, pp. 173–182.

30. P. Kyasanur and N. Vaidya, "Selfish MAC Layer Misbehavior in Wireless Networks," in *IEEE Trans. Mobile Comput.*, vol. 4, no. 5, pp. 502–516, 2005.

31. M. Luo, T. Peng and C. Leckie "CPU-Based DoS Attacks Agaisnt SIP Servers," in *Proc. IEEE Network Operations and Management Symposium*, 2008, pp. 41–48.

32. S. Niccolini, R. Garroppo, S. Giordano, G. Risi and S. Ventura, "SIP Intrusion Detection and Prevention: Recommendations and Prototype Implementation," in *Proc. 1st IEEE Workshop on VoIP Management and Security*, 2006, pp. 47–52.

33. P. Nuggehalli, M. Sarkar and R. Rao, "QoS and Selfish Users: A MAC Layer Perspective," in *Proc. IEEE GLOBECOM*, 2007, pp. 4719–4723.

34. K. Pelechrinis, G. Yan, S. Eidenbenz and V. Krishnamurthy, "Detecting Selfish Exploitation of Carrier Sensing in 802.11 Networks," in *Proc. IEEE INFOCOM*, 2009, pp. 657–665.

35. S. Radosavac, J. S. Baras and I. Koutsopoulos, "A Framework for MAC Protocol Misbehavior Detection in Wireless Networks," in *Proc. ACM Workshop on Wireless Security*, 2005, pp. 33–42.

36. S. Radosavac, G. Moustakides, J. Baras and I. Koutsopoulos, "An Analytic Framework for Modeling and Detecting Access Layer Misbehavior in Wireless Networks," in *ACM Trans. Information and Systems Security*, vol. 11, no. 4, article no. 19, Jul. 2008.

37. M. Raya, J. Hubaux and I. Aad, "DOMINO: A System to Detect Greedy Behavior in IEEE 802.11 Hotspots," in *Proc. ACM MobiSys*, 2004.

38. Y. Rong, S. Lee and H. Choi, "Detecting Stations Cheating on Backoff Rules in 802.11 Networks using Sequential Analysis," in *Proc. IEEE INFOCOM*, 2006, pp. 1–13.

39. J. Rosenberg, H. Schulzrinne and G. Camarillo, "SIP: Session Initiation Protocol," IETF RFC 3261, Jun. 2002.

40. S. Sawda and P. Urien "SIP Security Attacks and Solutions: A State-of-the-Art Review," in *Proc. IEEE 2nd Information and Communication Technologies*, 2006, pp. 3187–3191.

41. R. Schweller, Z. Li, Y. Chen, Y. Gao, A. Gupta, Y. Zhang, P. Dinda, M. Kao and G. Memik "Reverse Hashing for High-Speed Network Monitoring: Algorithms, Evaluation, and Applications" in *Proc. IEEE INFOCOM*, 2006.

42. H. Sengar, H. Wang, D. Wijesekera and S. Jajodia, "Fast Detection of Denial-of-Service Attacks on IP Telephony," in *Proc. 14th IEEE International Workshop on Quality of Service*, 2006, pp. 199–208.

43. H. Sengar, D. Wijesekera, H. Wang and S. Jajodia, "VoIP Intrusion Detection Through Interacting Protocol State Machines," in *Proc. IEEE International Conference on Dependable Systems and Networks*, 2006.

44. H. Sengar, H. Wang, D. Wijesekera and S. Jajodia, "Detecting VoIP Floods Using the Hellinger Distance," *IEEE Trans. Parallel Distrib. Syst.*, vol. 19, no. 6, pp. 794–805, Jun. 2008.

45. D. Sisalem, J. Kuthan and S. Ehlert, "Denial of Service Attacks Targeting a SIP VoIP Infrastructure: Attack Scenarios and Prevention Mechanisms," *IEEE Network*, vol. 20, no. 5, pp. 26–31, 2006.

46. M. Stephens, "EDF Statistics for Goodness of Fit and Some Comparisons," *Journal of the American Statistical Association*, vol. 69, pp. 730–737, 1974.

47. The MAdWiFi Driver, [Online.] Available: http://madwifi-project.org/.

48. A. Toledo and X. Wang, "A Robust Kolmogorov-Smirnov Detector for Misbehavior IEEE 802.11 DCF," in *Proc. IEEE ICC*, 2007, pp. 1564–1569.

49. A. Toledo and X. Wang, "Robust Detection of Selfish Misbehavior in Wireless Networks," in *IEEE J. Sel. Areas Commun.*, vol. 25, no. 6, pp. 1124–1134, Aug. 2007.

50. A. Toledo and X. Wang, "Detecting MAC Layer Collision Abnormalities in CSMA/CA Wireless Networks," in *Proc. IEEE ICC*, 2008, pp. 1598–1604.

51. A. Toledo and X. Wang, "Robust Detection of MAC Layer Denial-of-Service Attacks in CSMA/CA Wireless Networks," in *IEEE Trans. Information Forensics and Security*, vol. 3, no. 3, pp. 347–358, Sept. 2008.

52. VoIPSA, "VoIP Security and Privacy Threat Taxonomy," Public Release 1.0, 2005.

53. S. Vuong and Y. Bai, "A Survey of VoIP Intrusion and Intrusion Detection System," in *Proc. IEEE 6th International Conference on Advanced Communication Technology*, 2004, pp. 317–322.

54. G. Yang and L. Le Cam, *Asymptotics in Statistics: Some Basic Concepts*, second edition, Wiley, Mar. 2006.

55. H. Zhai, X. Chen and Y. Fang, "How Well Can the IEEE 802.11 Wireless LAN Support Quality of Service?" in *IEEE Trans. Wireless Communications*, vol. 4, no. 6, pp. 3084–3094, Nov. 2005.

56. "IEEE Standard for Wireless LAN-Medium Access Control and Physical Layer Specification," P802.11, 1999.

Chapter 2
Real-Time Misbehavior Detection in IEEE 802.11^{TM}: An Analytical Approach

Abstract In this chapter, we address the selfish misbehavior in the IEEE 802.11^{TM} based wireless network. After a brief description on selfish misbehavior in 802.11^{TM}, we first design a real-time backoff misbehavior detector, termed as the fair share detector (FS detector), which exploits the non-parametric cumulative sum (CUSUM) test to quickly find a selfish malicious node without any a priori knowledge of the statistics of the selfish misbehavior. We then develop a Markov chain based analytical model to systematically study the performance of the FS detector. Based on the analytical model, we can quantitatively compute the system configuration parameters for guaranteed performance in terms of average false positive rate, average detection delay and missed detection ratio under a detection delay constraint. We present simulation results to confirm the accuracy of our theoretical analysis as well as demonstrate the performance of the FS detector.

2.1 Selfish Misbehavior in 802.11^{TM}

2.1.1 IEEE 802.11^{TM} DCF

There are two major functions in the IEEE 802.11^{TM} protocols: the point coordination function (PCF) and the distributed coordination function (DCF). The PCF is a centralized function and is an optional feature in 802.11^{TM}. In this book, our concentration is the more widely used DCF which operates in a distributed manner. In the DCF, every node contends for access to the wireless medium following the CSMA/CA function [1]. When a node attempts to transmit a packet, it needs to sense the medium idle for a specified time. The time is divided into slots, and a node can only transmit at the beginning of a slot time. If the medium is not idle, the node will enter a backoff stage and defer the transmission according to a timer before attempting the next transmission. This backoff timer is a random value uniformly selected from a set $\{0, 1, \ldots, CW_{min} - 1\}$, where CW_{min} is called the minimum contention window with a standard value of 32. The timer will decrease if the medium is continuously sensed idle and freeze whenever the medium is sensed busy. After the timer reaches 0 the node will attempt another transmission. Each unsuccessful transmission due to reasons such as collisions or lost of ACK messages from the reception node will result in a doubled contention window size until it reaches the maximum contention window $CW_{max} = 2^m CW_{min}$, where m is called the maximum backoff stage with a standard value of 5. This operation is also referred to as the binary exponential backoff scheme. After a successful transmission, the node will reset the contention window to the minimum value CW_{min} and continue sensing the medium if it has more packets to transmit.

2.1.2 Backoff Selfish Misbehavior in IEEE 802.11™ DCF

As a distributed contention-based protocol, the DCF assumes that every node in the network operates in accordance with the protocol rules as described above to obtain a fair share of the wireless medium. However, a node which has the smallest backoff timer will obviously be favored by the protocol as it can always obtain more chances to transmit while other nodes are still in the backoff stage. Since there is no central controlling unit which assigns the backoff timer for each node, a malicious node can continuously choose a small backoff timer and then gain significant advantages in channel access probability over others. Moreover, because the increased transmission probability of the malicious node causes more collisions, normal nodes are forced to further exponentially defer their transmissions as they operate according to the protocol, which results in the malicious node gaining more advantages. The backoff misbehavior can drastically decrease the transmission probability of normal nodes and subsequently severely downgrade their throughput. In some extreme case where a malicious node sets its own backoff timer to a very small constant value, it will lead to denial of service (DoS) of the whole network except for the malicious node itself. Thus, a detection scheme capable of quickly and accurately identifying the misbehaving malicious node is highly desired for the normal operation of an IEEE 802.11™ wireless network.

2.2 Fair Share Detector Design

We consider a saturated situation that a node always has data to send when the channel is available. Although a network in practice is not always saturated, the saturated scenario is of meaningful concern in the context of selfish misbehaving. If the network is lightly loaded, a misbehaving node will not impact much the throughput of normal ones. When the network is close to full utilization, the data buffer in every node has a very small probability to be empty, where the saturated model is a good approximation.

2.2.1 The Observation Measure

Consider a tagged node v. In our detection system, the *observation measure* is an indicator of whether a successful transmission over the network belongs to the tagged node v, denoted as I^v. We take the popular modeling technique [1] that each node independently accesses an idle channel for transmission with a probability determined by its contention window size. If we use q_s^v to denote the probability that a successful transmission over the network is from node v, the probability distribution of I^v is given by

$$P\{I^v = k\} = \begin{cases} q_s^v & \text{if} \quad k = 1, \\ 1 - q_s^v & \text{if} \quad k = 0. \end{cases} \tag{2.1}$$

In a normal situation that every node uses the same contention window size and follows the 802.11^{TM} DCF model, it can be seen that $q_s^v = \frac{1}{N}$ under the independent channel access assumption and fair channel sharing, given N nodes in the network. If node v is a malicious node taking a smaller contention window size, it will achieve a q_s^v larger than $\frac{1}{N}$ and thus a larger portion of the network throughput. In Sect. 2.4, we will present how to calculate q_s^v given the contention window size. The distribution of I^v in (2.1) is the basis to establish our analytical model.

Remark 2.1: In an 802.11^{TM} network, a node that has just accomplished a successful transmission will have advantages in grabbing the channel for next transmission in a short period [3]. This is referred to as *short-term unfairness* and is inherent to the 802.11^{TM} backoff mechanism. Such an issue implies correlations among the channel accesses, which may impact the accuracy of (2.1) to model the successful transmission of the tagged node based on the assumption of independent channel access. In our preliminary work [5], we apply a shuffling mechanism to observation samples to mitigate the impact of short-term unfairness. In Sect. 2.5, we will show with detailed analysis that the FS detector is inherently robust against short-term unfairness, and the detection based on (2.1) does give accurate decisions. The fairness issue also exists when both user datagram protocol (UDP) and transmission control protocol (TCP) traffic flows exist in the network, where the TCP traffic tends to be overwhelmed by UDP traffic due to its congestion control mechanism. In Sect. 2.5, we will also discuss how to apply the FS detector with a robust performance when both UDP and TCP traffic flows exist in the network.

2.2.2 Fair Share Detector

Let $\{I_n, n = 0, 1, \ldots.\}$ be the sequence of sample values of I^v, observed each time a successful transmission appears on the channel. Here, we drop the superscript v for easier presentation considering the clear context. Let N denote the number of nodes existing in the network. Suppose that the initial value of our detector X_n is 0. When the current successful transmission over the network is from the tagged node, i.e., $I_n = 1$, we increase X_n by $N - 1$; otherwise, when the transmission is from any non-tagged node, i.e., $I_n = 0$, we decrease X_n by 1 until it reaches 0. The intuition of this design is as follows: In the normal situation where every node follows the 802.11^{TM} DCF model, each node roughly takes turn to transmit; the increase of X_n caused by one successful transmission from the tagged node can then be equally offset by the successful transmissions from other $N - 1$ non-tagged nodes. Thus in the normal situation, the detector X_n will fluctuate around a low value close to zero. On the other hand, when the tagged node turns to misbehave and obtain more chances to transmit, it is not difficult to see that X_n is going to quickly accumulate to a large positive value.

The behavior of the FS detector can be mathematically described as

$$X_n = (X_{n-1} + (NI_n - 1))^+$$
$$X_1 = 0 \tag{2.2}$$

where $(x)^+ = x$ if $x \geq 0$ or 0 otherwise. We can see that (2.2) is actually in the form of a non-parametric CUSUM detector [2].

Let h be the detection threshold, then the decision rule of the detector in step n is

$$\delta_n = \begin{cases} 1 & \text{if} \quad X_n \geq h \\ 0 & \text{if} \quad X_n < h \end{cases} \tag{2.3}$$

where δ_n is also an indicator function of whether the detection event happens or not. The detector value X_n will be reset back to 0 as soon as it exceeds the threshold and the detection procedure starts over again.

2.3 Markov Chain Based Analytical Model

Consider the sequence $\{X_n\}$ as a discrete random process, which takes values from a finite set $A = \{0, 1, 2, \ldots, h\}$. The process is said to be in state j at time n if $X_n = j$ and in state i at time $n - 1$ if $X_{n-1} = i$, where $i, j \in A$. The transition between the states happens when a successful transmission over the network is observed. According to (2.2), the current state X_n depends only on the state X_{n-1} and is independent of any other previous states, where the transition probability is

$$P_{ij} = P\{X_n = j | X_{n-1} = i\}. \tag{2.4}$$

Thus the random process $\{X_n\}$ satisfies the Markov property and can be modeled as a discrete-time Markov chain.

Given the decision threshold h, the Markov chain is then described by a $(h + 1) \times (h + 1)$ transition probability matrix as

$$\mathbf{P} = \begin{pmatrix} P_{00} & P_{01} & P_{02} & \cdots & P_{0h} \\ P_{10} & P_{11} & P_{12} & \cdots & P_{1h} \\ \vdots & \vdots & \vdots & \ddots & \vdots \\ P_{h0} & P_{h1} & P_{h2} & \cdots & P_{hh} \end{pmatrix}.$$

This transition probability matrix can be divided into three distinct groups based on the operation of the FS detector.

Group 1 consists of P_{ij} for $i = 0$ and $j \in [0, h]$, with values

$$P_{0j} = \begin{cases} P\{I_n = 0\} & \text{if } j = 0, \\ P\{I_n = 1\} & \text{if } j = N - 1 \text{ and} \\ & N - 1 \leq h, \\ P\{I_n = 1\} & \text{if } j = h \text{ and} \\ & N - 1 > h, \\ 0 & \text{otherwise.} \end{cases} \tag{2.5}$$

This group is related to the transitions from state 0 to other states. According to the state transition equation (2.2), the detector variable X_n jumps out of state 0 only when the observed successful transmission is from the tagged node, that is, $I_n = 1$. Further, X_n makes a transition to either $N - 1$ or h depending on whether $N - 1$ is greater than h or not. Note that the state h in fact incorporates all possible states $X_n \geq h$, as the detector will raise an alarm when the state hits h.

Group 2 consists of P_{ij} for $i \in [1, h - 1]$ and $j \in [0, h]$, with values

$$P_{ij} = \begin{cases} P\{I_n = 0\} & \text{if } j = i - 1, \\ P\{I_n = 1\} & \text{if } j = i + N - 1 \text{ and} \\ & i + N - 1 \leq h, \\ P\{I_n = 1\} & \text{if } j = h \text{ and} \\ & i + N - 1 > h, \\ 0 & \text{otherwise.} \end{cases} \tag{2.6}$$

This group describes the typical behavior of the detector. The state can transit to left (i.e., to a smaller value) when $I_n = 0$ or to right (i.e., to a larger value) when $I_n = 1$, according to the state transition equation (2.2).

Finally, group 3 consists of P_{ij} for $i = h$ and $j \in [0, h]$, with values

$$P_{hj} = \begin{cases} 1 & \text{if } j = 0, \\ 0 & \text{otherwise.} \end{cases} \tag{2.7}$$

This group is related to the transitions out of state h. Since the detector value will be reset to 0 as soon as it reaches or exceeds h, $P_{h0} = 1$.

2.4 Theoretical Performance Analysis

In this section, we conduct theoretical performance analysis of the FS detector based on the Markov chain model in terms of the three fundamental metrics to change detection: average false positive rate, average detection delay, and missed detection ratio under a detection delay bound. Then we show how we can configure the system parameters to achieve guaranteed performance. We also analyze the performance of the detector when the number of nodes is varying, which is a typical scenario in the 802.11^{TM} based wireless networks.

2.4.1 Average False Positive Rate

The average false positive rate P_{fp} is the rate that the detector value X_n hits state h given the fact that there is no node in the network misbehaving. According to the theory on the discrete-time Markov chain, such a rate is equal to the steady-state probability that the Markov chain describing the FS detector stays at h in the normal condition.

In the normal condition with a fair share of the channel access, we have $q_s^v = \frac{1}{N}$ for a tagged node. We can calculate the distribution of I_n according to (2.1), and further obtain the transition probabilities matrix $\mathbf{P}$ according to (2.5)–(2.7).

Let $(\pi_0, \ldots, \pi_h)$ denote the steady state probabilities of the Markov chain, which can be solved from the equations

$$\pi_j = \sum_{i=0}^{h} \pi_i P_{ij}, \quad j \in \{0, \ldots, h\}, \tag{2.8}$$

$$\sum_{j=0}^{h} \pi_j = 1. \tag{2.9}$$

Then we can get the average false positive rate

$$P_{fp} = \pi_h. \tag{2.10}$$

The analytical result (2.10) allows us to numerically examine the impact of the fundamental parameter h on the average false positive rate P_{fp} of the FS detector. As an example, we compute the results for a network with $N = 10$ nodes, and the results are illustrated in Fig. 2.1. From the figure, we can observe that a larger h yields a smaller false positive rate, as expected.

2.4.2 Average Detection Delay

In this subsection, we analyze the average detection delay denoted as $E[T_D]$, which is the average number of samples observed from the moment that the tagged node

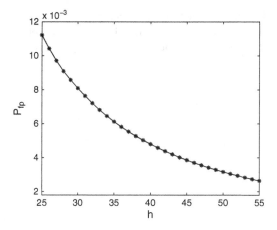

Fig. 2.1 Average false positive rate

starts to misbehave until the misbehavior is detected. With the Markov chain under the abnormal condition (abnormal Markov chain), $E[T_D]$ can be computed as the expected number of transitions required for the state variable to hit state h, starting from the moment when the misbehavior starts. To carry out the analysis, we need to find the transition probabilities of the abnormal Markov chain and determine the initial state of the FS detector when the misbehavior starts.

2.4.2.1 Transition Probabilities Under the Misbehavior

We consider a network consisting of two classes of nodes. Class 1 includes the one misbehaving node with a small minimum contention window CW_{min} denoted as W_1, and class 0 includes all the normal nodes with the standard minimum contention window denoted as W_0. According to the classic modeling approach for the 802.11™ DCF [1], we consider that each node independently accesses an idle channel for transmission. Let p_t^i denote the probability that a class i ($i \in 0, 1$) node transmits at a random time slot and p_c^i denote the collision probability of a class i node. Also recall that N is the number of nodes and m is the maximum backoff stage. According to [1], we have the following equations:

$$
\begin{cases}
p_t^0 = \dfrac{2(1-2p_c^0)}{(1-2p_c^0)(W_0+1)+p_c^0 W_0(1-(2p_c^0)^m)} \\[2ex]
p_t^1 = \dfrac{2(1-2p_c^1)}{(1-2p_c^1)(W_1+1)+p_c^1 W_1(1-(2p_c^1)^m)} \\[2ex]
p_c^0 = 1-(1-p_t^1)(1-p_t^0)^{N-2} \\[1ex]
p_c^1 = 1-(1-p_t^0)^{N-1}
\end{cases}
\tag{2.11}
$$

from which the four parameters p_t^0, p_t^1, p_c^0 and p_c^1 can be solved.

Note that a node can get a successful transmission under the circumstance that there is no collision while the node transmits. Thus from the solutions of (2.11), we can obtain the probability that a node gets a successful transmission at a random time slot:

$$p_s^0 = p_t^0(1 - p_c^0), \tag{2.12}$$

$$p_s^1 = p_t^1(1 - p_c^1). \tag{2.13}$$

We can then calculate the probability $\hat{q}_s$ that a successful transmission over the network is from the malicious node as (2.14):

$$\hat{q}_s = \frac{p_s^1}{p_s^1 + (N-1)p_s^0}. \tag{2.14}$$

Using $\hat{q}_s$ in (2.1), we can obtain the distribution of I_n for the misbehaving node; using such I_n distribution in (2.5)–(2.7), we can then compute the transition probability matrix $\hat{\mathbf{P}}$ for the abnormal Markov chain.

It is worth noting that although we only include two classes of nodes in the above analysis, the model of (2.11)–(2.14) can be easily extended to cases where multiple classes of misbehaving nodes with different intensities of misbehavior exist. This will enable us to analyze much more complicated misbehaving scenarios. We will discuss this issue in Sect. 2.4.4.

2.4.2.2 Initial States

A natural thought of the initial state of X_n is 0 when the misbehavior starts. However, this may not be the case; before a malicious node starts to misbehave, it can behave like a normal node and still affect X_n. Thus X_n can be initially at any state following the normal Markov chain except for state h, as we do not consider an already "alarmed" state as an initial state.

We can calculate the steady state probabilities of the normal Markov chain according to (2.8) and (2.9). Since we are interested in detection starting from an unalarmed state, under such a constraint the conditional initial state probabilities should be

$$\pi_i' = \frac{\pi_i}{\sum_{i=0}^{h-1} \pi_i} \quad \text{for } i \in \{0, \dots, h-1\}. \tag{2.15}$$

2.4.2.3 Average Detection Delay

As we have various initial states, the average detection delay $E[T_D]$ should be calculated as the weighted average of the expected numbers of transitions from every initial state to state h based on the transition probability matrix $\hat{\mathbf{P}}$ for the abnormal Markov chain.

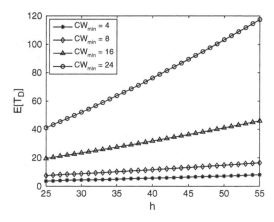

Fig. 2.2 Average detection delay

Let μ_{ih}, $i \in [0, h-1]$, denote the expected number of transitions for state i to state h. According to [4], the values of μ_{ih} can be solved from the equations

$$\mu_{ih} = 1 + \sum_{r \neq h} \hat{P}_{ir} \mu_{rh}, \quad i \in \{0, \ldots, h-1\} \tag{2.16}$$

where $\hat{P}_{ir}$ is the transition probability from state i to r of $\hat{\mathbf{P}}$. Based on the solutions of (2.15) and (2.16), we can obtain the average detection delay $E[T_D]$ as

$$E[T_D] = \sum_{i=0}^{h-1} \pi_i' \mu_{ih}. \tag{2.17}$$

The analytical result (2.17) allows us to numerically examine the impact of h on the average detection delay $E[T_D]$ of the FS detector. As an example, we compute the results for a network with $N = 10$ nodes, and the results are illustrated in Fig. 2.2. Specifically, Fig. 2.2 shows the analysis results under four misbehaving intensities $CW_{min} = 4$, $CW_{min} = 8$, $CW_{min} = 16$ and $CW_{min} = 24$. As we expect, the figure illustrates that more intense misbehavior leads to a shorter detection delay. Also, we observe that a smaller h yields better performance in average detection delay.

2.4.3 Missed Detection Ratio

In this subsection, we discuss the missed detection ratio, denoted as P_{md}. The FS detector exploits the non-parametric CUSUM test. The missed detection ratio is not often considered in the context of CUSUM test due to its "non-stop until detection" property. We however examine P_{md} under a given detection delay constraint D, which is of importance regarding real-time detection.

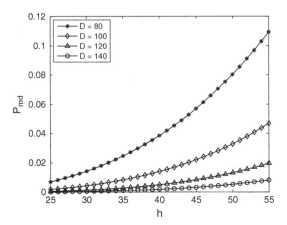

Fig. 2.3 Missed detection ratio

The detection event happens only when X_n hits state h. Thus the missed detection ratio P_{md} under the delay constraint D is the summation of the probabilities of X_n staying at a state other than h at time D. With the transition probability matrix $\hat{\mathbf{P}}$, the missed detection ratio can be computed in an iterative manner. Let the row vector $\mathbf{P}(j) = [P_0(j), \cdots, P_h(j)]$ denote the probabilities of the state variable at step j with $0 \leq j \leq D$. The computation starts from the initial states given in (2.15), setting

$$P_i(0) = \pi_i' \quad \text{for } i \in \{0, \ldots, h-1\}, \tag{2.18}$$
$$P_h(0) = 0. \tag{2.19}$$

At each transition step $j \in [0, D-1]$, the state probabilities are updated as

$$\mathbf{P}(j) = \mathbf{P}(j-1) \cdot \hat{\mathbf{P}}, \tag{2.20}$$
$$P_h(j) = 0. \tag{2.21}$$

At each step, $P_h(j)$ is set to 0 for next step computation because we are interested in the missed detection cases. The missed detection ratio under the delay bound constraint D can be obtained as

$$P_{md} = \sum_{i=0}^{h-1} P_i(D). \tag{2.22}$$

Figure 2.3 demonstrates the missed detection ratios P_{md} of our analysis under the delay constraints $D = 80$, $D = 100$, $D = 120$ and $D = 140$, for a misbehaving node with the moderate misbehavior of $CW_{min} = 16$. We observe that the larger the delay constraint is, the lower the missed detection ratio will be. In other words, the probability of detection increases with a cost of longer delay. Also, a smaller detection threshold h yields a lower missed detection ratio.

2.4.4 Discussion on Detection of Multiple Misbehaving Nodes

Our analytical model can be extended to the cases where multiple classes of malicious nodes with different intensities of misbehavior exist. The key to the analysis is to obtain the abnormal Markov chain, which in fact is determined by the probability that a successful transmission over the network is from the tagged malicious node.

Consider a network of N nodes, k of which are malicious and the rest are normal. Suppose that malicious node i sets its minimum contention window CW_{min} as W_i and all the $N - k$ normal nodes use the standard minimum contention window denoted as W_0. We can expand (2.11) to have the following equations:

$$
\begin{cases}
p_t^0 = \dfrac{2(1 - 2p_c^0)}{(1 - 2p_c^0)(W_0 + 1) + p_c^0 W_0 (1 - (2p_c^0)^m)} \\[2mm]
\quad \vdots \\[2mm]
p_t^k = \dfrac{2(1 - 2p_c^k)}{(1 - 2p_c^k)(W_k + 1) + p_c^k W_k (1 - (2p_c^k)^m)} \\[2mm]
p_c^0 = 1 - \displaystyle\prod_{i=1}^{k}(1 - p_t^i)(1 - p_t^0)^{N-k-1} \\[2mm]
\quad \vdots \\[2mm]
p_c^k = 1 - \displaystyle\prod_{i=1}^{k-1}(1 - p_t^i)(1 - p_t^0)^{N-k}
\end{cases}
\tag{2.23}
$$

From the solutions of (2.23), we can obtain the probability that a node gets a successful transmission at a random time slot:

$$
p_s^0 = p_t^0(1 - p_c^0),
\tag{2.24}
$$

$$
\vdots
$$

$$
p_s^k = p_t^k(1 - p_c^k).
\tag{2.25}
$$

Then we can calculate the probability q_s^l that a successful transmission over the network is from the tagged malicious node l with $CW_{min} = W_l$ as

$$
\hat{q}_s^l = \frac{p_s^l}{\sum_{i=1}^{k} p_s^i + (N - k)p_s^0}.
\tag{2.26}
$$

Using q_s^l in (2.26), we can obtain the transition probability matrix of the abnormal Markov chain $\hat{\mathbf{P}}^l$; using $\hat{\mathbf{P}}^l$ and initial states of the detector when misbehavior starts, which are determined in the same way as in Sect. 2.4.2.2, we can analyze average detection delay and missed detection ratio for the tagged malicious node l accordingly.

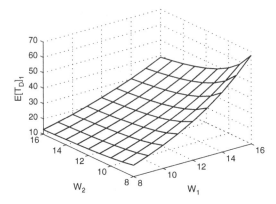

Fig. 2.4 Average detection delay under two misbehaving nodes

We consider an example that there are two misbehaving nodes in a network of 10 nodes, one setting its minimum contention window as W_1 and the other as W_2. Figure 2.4 plots the average detection delays to identify the misbehaving node 1, denoted as $E[T_D]_1$, under different misbehaving intensity pairs (W_1, W_2). Note that even in this simple case the two malicious nodes are competing with each other. There is a trade-off between the two nodes. Certainly it takes longer to detect one malicious node if the other chooses more intense misbehavior. It will be an interesting problem to determine how the multiple malicious nodes can find certain misbehaving strategies to collaboratively maximize their collective benefit from the network throughput while avoiding being detected as long as possible. In next chapter, we will also carry out in-depth studies of the scenario with multiple misbehaving nodes when our more advanced adaptive detector is considered.

2.4.5 System Configuration Under Performance Constraints

The above theoretical analysis provides us a guideline to configure the system parameter h for guaranteed performance in a target scenario. For each performance metric, we can obtain the feasible ranges of h to satisfy the performance constraints. With the intersection of the parameter ranges under all the constraints, a proper configuration of h can be obtained to meet the performance requirements of all the metrics. Moreover, once we determine the configuration parameter, we can explicitly estimate the performance measures given a misbehaving scenario. In practice, as we do not have a priori knowledge of the misbehavior, the analytical model allows us to conservatively configure the system so that even the misbehavior with a low intensity can be detected with good performance. For example, if we select $h = 40$ for a network with $N = 10$, our analytical model indicates that, even for the moderate misbehavior with $CW_{min} = 16$, we can target a high level of performance with the average false positive rate of 0.005, the average detection delay of 31.8357

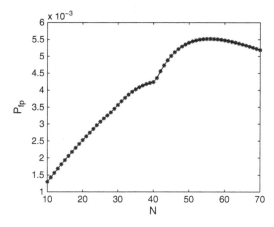

Fig. 2.5 Impact of network size change on average false positive rates at $h = 80$

samples, and the missed detection ratio of 0.0141 with the delay constraint $D = 100$. In Sect. 2.5, we will use simulation results to demonstrate that our target performance measures are indeed achievable.

2.4.6 Detection Performance Against Network Size Change

In an 802.11^{TM} based wireless network, it is typical that nodes are mobile and thus the number of nodes (i.e., the network size) changes from time to time. The FS detector is robust against such a scenario. As we directly include the number of nodes N in the detector design, when N changes, the detector can adjust and respond in real time.

Figure 2.5 shows the average false positive rates P_{fp} of the detector versus the number of nodes N, at $h = 80$. The threshold h is intentionally set to be greater than the maximum number of nodes to avoid alarm being triggered by just one successful transmission from the tagged node. As shown in Fig. 2.5, there is a dent on the curve at $N = 40$ and P_{fp} has a sharper increase when N gets greater than 40. This is because, when $N \leq 40$, at least three or more consecutive successful transmissions from the tagged node are needed to drive X_n to h from an initial state of 0, raising a false alarm; however, when $41 \leq N \leq 70$, it will take only two consecutive transmissions to reach h, which largely increases the possibility of false positive. Furthermore, note that P_{fp} does not monotonically increase with N and has an upper bound of $P_{fp} = 0.0055$. The explanation is that, when the number of nodes contending for the channel becomes larger, the transmissions from a tagged node are more likely to be interrupted by transmissions from those non-tagged nodes, and the accumulation of the detector X_n will be more aggressively offset by such non-tagged nodes, thus resulting in a smaller P_{fp}. If the target performance of $P_{fp} \leq 0.0055$ is allowed,

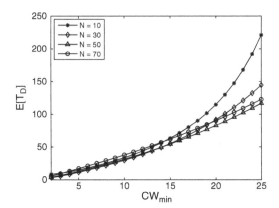

Fig. 2.6 Impact of network size change on average detection delays at $h = 80$

we can see that the configuration $h = 80$ satisfies the false positive performance requirement even when N changes dynamically in a wide range. Note that a typical 802.11^{TM} based wireless local area network covers up to tens of users.

Fixing $h = 80$, we now investigate the average detection delay $E[T_D]$ of the detector for different misbehavior intensities, indicated by the CW_{min} value of a misbehaving node, with results shown in Fig. 2.6. The misbehavior intensities with $CW_{min} > 25$ are not included in our discussion, as their effects are minimal. Practically, a misbehaving node needs to choose more intense misbehavior, e.g., $CW_{min} \leq 16$, to gain more benefits from the network throughput. From Fig. 2.6, we see that for misbehavior in this range, the change of N does not affect $E[T_D]$ much. The reason is that, when a misbehaving node grabs the channel, very likely it will consecutively send a certain number of packets, driving the detector to hit the threshold. For a smaller value of N, it may just take a couple of more samples for the detector to hit the threshold (note that each transmission from the tagged node increases the detector state by $N - 1$), which only slightly increases the detection delay. With less intense misbehavior ($16 < CW_{min} \leq 25$), we do observe obviously larger detection delays for a small N. The reason is that, when the misbehaving intensity is low, the accumulation procedure of X_n is more often to be offset by transmissions from those non-tagged normal nodes; for a small N, it will take even more samples from the misbehaving node to raise the alarm, leading to a longer detection delay.

It is noteworthy that the relationship between N and detection delay in Fig. 2.6 is not rigorously monotonic. Such phenomenon is due to two contradicting factors: Given the CW_{min}, the misbehaving node will get less transmissions when N gets larger and thus less chances to accumulate X_n, potentially increasing the detection delay; the increase of X_n (by a value of $N - 1$) caused by one transmission from the misbehaving node however becomes larger too, potentially decreasing the detection delay. In summary, the results in Fig. 2.6 again demonstrate that the FS detector with a fixed threshold (larger than N) has a robust performance for a typical misbehaving scenario, even when the number of nodes in the network changes.

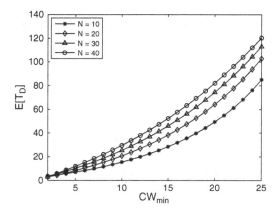

Fig. 2.7 Impact of network size change on average detection delays at $P_{fp} = 0.005$

In a situation where a fixed constraint on P_{fp} is imposed, we can dynamically calculate the h value corresponding to a certain N through the analytical model. Further, if we do the calculation beforehand and maintain a table of "h versus N" values under the given P_{fp} constraint, we can quickly adjust h as soon as changes on N are observed. Figure 2.7 shows the average detection delays $E[T_D]$ of the detector for different misbehaving intensities, given a false positive constraint as $P_{fp} = 0.005$. Similar to Figs. 2.6 and 2.7 shows that the detection delays under different N are similar when the misbehavior is very intense. Under a lower misbehaving intensity (i.e., a larger CW_{min}), the detection delays increase more obviously with the number of nodes, because a larger threshold h is required for a larger N to meet the false positive requirement. However, the delay increase is not dramatic. Even for $CW_{min} = 25$ and $N = 40$, it only takes about 120 successful transmissions over the whole network to detect the misbehavior.

2.4.7 Comparison with the Original CUSUM Detector

In order to show how we have improved in real-time misbehavior detection, we compare the FS detector to the detector developed in our preliminary work [5], referred to as the "original CUSUM detector" for convenience. The observation measure of the original CUSUM detector is the number of successful transmissions of the tagged node in every M successful transmissions over the whole network. It means getting one observation sample for the original CUSUM detector requires M successful transmissions, whereas the FS detector will update state upon every successful transmission over the network. Also, M needs to be at least as large as the number of nodes N and linearly increase with N to fairly count transmissions from each node. Moreover, besides h, there is another parameter u in the original detector design, which is the upper bound of the observation measure's

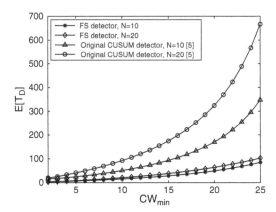

Fig. 2.8 Comparison with the original CUSUM detector in [5] at $P_{fp} = 0.005$

expectation. To determine a proper u, we need to take into account both the sample size M and the number of nodes N, adding the complexity of the detection system. In the FS detector, u is not present, which leads to one less parameter impacting the detection performance and thus makes parameter configuration much simpler.

Figure 2.8 shows the average detection delays of the two detectors for different misbehavior intensities under the same false positive constraint of $P_{fp} = 0.005$. Here we consider the cases of $N = 10$ and $N = 20$. For the FS detector, given the P_{fp} and N, the threshold h can be determined from the analytical model. With h, the detection delay for a given misbehaving intensity can then be calculated and plotted in Fig. 2.8. We intentionally configure the original CUSUM detector for a small detection delay so the advantage of the FS detector can be demonstrated more convincingly. The sample size M for the original CUSUM detector is set to its minimum value N (i.e., 10 and 20 for the two cases considered, respectively) in order to minimize the impact of the observation window size on the detection delay. With such an observation window selection, on average one successful transmission from each node can be expected in each window, i.e., $u = 1$. Given the P_{fp}, N and u, the parameter h can then be determined from the analytical model in [5]. With h and u, the detection delay for a given misbehaving intensity with the original CUSUM detector can be calculated and plotted in Fig. 2.8.

As shown in Fig. 2.8, for the same N, the FS detector shows clear advantages over the original CUSUM detector, especially when the misbehavior becomes less intense. Observing the delays of the original CUSUM detector, we can see that the delays with $N = 20$ are roughly two times of the delays with $N = 10$ for almost all the misbehavior intensities. The fact clearly indicates the impact of the observation window size on detection delay in the original CUSUM detector. Another advantage of the FS detector is that its detection delay curves are quite flat against the misbehaving intensity and not much impacted by the network size N, showing very robust performance.

2.5 Simulation Results

2.5.1 Simulation Setup

We establish an 802.11^{TM} DCF based wireless network consisting of 10 competing nodes ($N = 10$) and an access point (AP) through ns-2 [6] simulation. We first consider that the network works under the saturated condition and every node sends packets with UDP towards the AP. Then we include the TCP traffic in our simulation to further analyze the performance of the FS detector in more general scenarios. The AP also acts as the detection agent which monitors the transmissions from every competing node with a separate FS detector. The nodes are located close enough to sense the transmissions from each other and thus avoid the hidden terminal problem. There is 1 misbehaving node among the 10 competing nodes, which accesses the wireless channel using the binary exponential backoff scheme but can manipulate its minimum contention window CW_{min} to any value between 1 and 32.

Due to the conflicting nature of the three performance metrics (average false positive rate, average detection delay, and missed detection ratio), it can be difficult to find the system configuration parameter that achieves best performance at all fronts. Using our analytical model, we find that, for $N = 10$, setting the detection threshold $h = 40$ can achieve a good tradeoff among all the metrics (referring to Sect. 2.4.5). Therefore in our simulation, if not specified, we set $h = 40$ to further evaluate the performance of our detection.

2.5.2 Robustness Against Short-Term Unfairness

In an 802.11^{TM} network, a node that has just accomplished a successful transmission will have advantages in grabbing the channel for next transmission in a short period [3]. This is referred to as *short-term unfairness* and is inherent to the 802.11^{TM} backoff mechanism. Such an issue implies correlations among the channel accesses, which impact the accuracy of the transition probability calculation based on the assumption of independent channel access. The system configuration based on an inaccurate model can lead to inaccurate detection results. In this section, we study how the short-term unfairness affects the performance of our detector.

We first examine the impact of short-term unfairness on the distribution of the detector X_n under the normal traffic condition. In Fig. 2.9, we present the simulation results of the cumulative distribution function (CDF) of X_n, compared with the analytical CDF. Note that even though the analytical results are based on the independent model of (2.1), the two curves are still close to each other. We then examine the average false positive rate P_{fp} versus h, comparing the analytical results with the simulation results in Fig. 2.10. Again, despite a bigger gap when h is smaller, the P_{fp} curve obtained from simulations still largely resembles the analytical one. The observations show that our FS detector is robust against the impact of short-term unfairness.

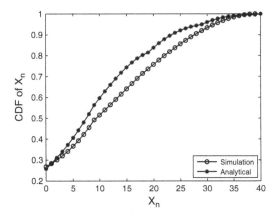

Fig. 2.9 CDF of X_n

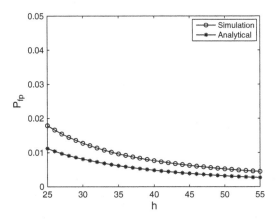

Fig. 2.10 Average false positive rate

We then obtain the average detection delays $E[T_D]$ under different misbehaving intensities. Figures 2.11 and 2.12 present both the simulation and analytical $E[T_D]$ curves versus h for $CW_{min} = 8$ and $CW_{min} = 16$, respectively. The closeness of the two curves in both cases again confirms the robustness of the FS detector against the short-term unfairness.

Technically, the FS detector by nature can mitigate the impact due to the short-term unfairness. In the normal situation, every node in the network has the same opportunity to experience a short period of advantages in transmissions. At a sampling moment, if the tagged node under observation is accessing the channel more aggressively due to short-term unfairness, it will increase the detector state value more aggressively according to (2.2), tending to be false positive. However, if at other sampling moments, those non-tagged nodes are accessing the channel more aggressively, it will in turn decrease the detector state value more aggressively and mitigate the false positive effect. Therefore, as an aggregate effect, the FS detector only degrades slightly in the false positive performance. In the misbehaving situa-

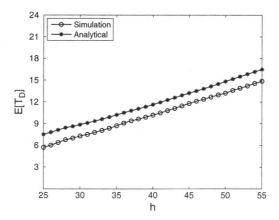

Fig. 2.11 Average detection delay with $CW_{min} = 8$

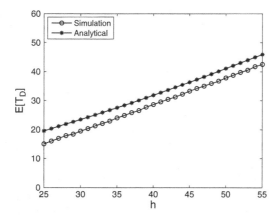

Fig. 2.12 Average detection delay with $CW_{min} = 16$

tion, extra channel access (in addition to that resulting from the backoff misbehavior) due to the short-term unfairness effect in fact benefits the misbehavior detection, with the detector being driven to hit the threshold h sooner, as shown in Figs. 2.11 and 2.12. We did design a shuffling mechanism based on the similar idea as that applied in [5] to address the impact of the short-term unfairness, and found that it would sacrifice a lot in detection delay to achieve just a moderate gain in mitigating false positive rate. Thus according to the theoretical and simulation investigations given above, we decide to apply the FS detector without an extra mechanism for the short-term unfairness issue.

2.5.3 Performance Guarantee

Given the configuration parameter $h = 40$, we compare the target performance measures with the simulation results under the same setting, shown in Table 2.1, to

Table 2.1 Comparison of analytical and simulation results with $N = 10$, $h = 40$, $D = 100$

	P_{fp}	$E[T_D]$	P_{md}
Analysis	0.005	31.8357	0.0141
Simulation	0.0076	28.5744	0.0255

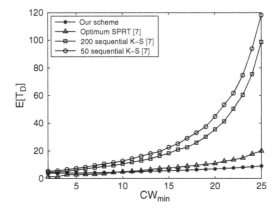

Fig. 2.13 Comparison with the detection schemes in [7]

examine whether the target performance is guaranteed. We can see that simulation results are very close to the target values in all three performance metrics. The small gap between the values is largely due to the variance in the observation samples; also the effect of the short-term unfairness is not 100 % overcome according to Figs. 2.9 and 2.10. Considering such a small gap, in practice we can on purpose select configuration parameters to conservatively provision the detection performance.

With the same parameter configuration as above, we compare our FS detector to the sequential K–S test and the optimal SPRT for 802.11^{TM} backoff misbehavior detection used in [7] in Fig. 2.13. The sample used in those solutions is collected every successful transmission of the tagged node, whereas in our scheme, the sample is collected every successful transmission from any node in the network. The average detection delays in terms of the number of successful transmissions from the tagged node for different detection schemes are compared in Fig. 2.13. For a fair comparison, we map our samples (the total number of successful transmissions over the network) to that used in [7]. For such a mapping, we only need to count the number of successful transmissions from the tagged node within the total successful transmissions. Also note that the desired false positive rate in [7] is fixed at $P_{fp} = 0.05$, which is one order larger than our target 0.005 as given in Table 2.1. Even with a much more strict constraint on P_{fp}, Fig. 2.13 shows that our detector has comparative detection delays against high intensities of backoff misbehavior and becomes superior to all other schemes as the misbehavior turns less intense.

It is interesting to discuss why our FS detector has better performance even than the optimal SPRT (when the misbehaving intensity is not high) in [7]. An optimal SPRT has the "optimal" performance only when the normal behavior distribution could be accurately obtained. However, to establish the normal behavior distribution, the detectors in [7] need to first estimate the collision probability over the 802.11^{TM} channel. In [7], there are two aspects of inaccuracy in estimating the collision probability, which degrade the performance of false positive rate and detection delay, respectively.

The first aspect of inaccuracy in [7] is that the collision probability is estimated from only tens of samples, over which the variance may lead to overestimating the collision probability. The behavior monitored by the detector is the idle time between consecutive successful transmissions; an overestimated collision probability will lead to an overestimated idle time (longer than its real value). With such an estimation error by the detector, a normal idle time observed will appear smaller than the "thought-to-be" normal behavior and thus misunderstood as misbehaving. That is, the overestimation of the collision probability leads to a higher false positive rate.

The second aspect of inaccuracy is that, according to the IEEE 802.11^{TM} model, a conditional collision probability (given that the tagged node is sending a packet) should be used to characterize the backoff procedure and further estimate the distribution of the idle time between consecutive successful transmissions. The study in [7] however uses an unconditional collision probability estimated over all nodes to approximate the conditional one. Regarding the misbehaving node, the unconditional collision probability will be an underestimate of the conditional one. The conditional collision probability associated with the tagged misbehaving node is determined by transmissions from other normal nodes. When estimating with an unconditional collision probability, transmissions from the misbehaving node are also included in the estimation [7]; note that many transmissions from the misbehaving node will not experience collisions due to the misbehaving node's aggressive access to the channel. Thus, the collision probability will be obviously underestimated, which then makes the detector to underestimate the normal idle time between consecutive successful transmissions. Such underestimation of the normal model makes the misbehavior deviation less obvious and incurs a longer time for detection.

2.5.4 Performance with UDP and TCP Traffic

We also consider scenarios where TCP traffic exists in the network. Figure 2.14 shows the average detection delay of a misbehaving node versus the misbehaving intensity in a network of 10 nodes. The detection threshold $h = 40$. We compare the detection delays in the two scenarios that all the nodes send TCP traffic or saturated UDP traffic to the AP, respectively. As shown in Fig. 2.14, in most cases, the detection delay in the TCP scenario is larger than that in the UDP one, especially when the misbehavior is more intense. The reason is that TCP multiplicatively decreases the transmission rate upon a packet loss due to its congestion control mechanism; the impact of congestion control is more obvious in wireless networks where collisions are common. The congestion control mechanism by nature mitigates the

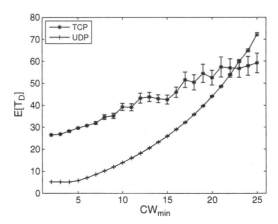

Fig. 2.14 Average detection delay under TCP traffic

selfish misbehavior. Aggressive transmissions will lead to more collisions, which in turn decreases the sending rate through the congestion control. Thus, it takes longer time to detect (compared to the UDP case) the misbehavior due to the mitigating effect of the congestion control. With a low misbehaving intensity ($CW_{min} > 20$), the congestion control effect applies more to the normal nodes, where the detection delay will be shorter than that in the UDP case. In Fig. 2.14, we also plot the 95 % confidence interval, measured from a large number of detections (in the order of 10^6), which can show that TCP congestion control brings a high degree of dynamics to the system.

A more general scenario would be a wireless network consisting of both TCP and UDP nodes.[1] There are three possible cases. (1) All the nodes have a normal behavior. In this case, the FS detector will have a high false positive rate to indicate a certain normal UDP node as a misbehaving node, since the throughput of UDP nodes will overwhelm those TCP ones. (2) A misbehaving node exists as a UDP node. Note that when both UDP and TCP flows exist, it is impossible for a TCP node to aggressively grab more throughput due to the congestion control. As a UDP misbehaving node will easily overwhelm those normal TCP nodes, the detection delay of a misbehaving node will be even shorter than that in an all-UDP case. (3) To avoid being detected, a smart misbehaving node may establish a TCP connection to the AP, but does not implement the congestion control mechanism (i.e., actually transmit according to UDP).

For robust detection performance in the complex scenario when both UDP and TCP traffic flows exist, we design a *dual-detector* implementation as shown in Fig. 2.15. FS detector 1 monitors traffic from all the nodes; if a detection event happens, we check whether the tagged node claims to use TCP or UDP. Then, if

---

[1] Without loss of generality, we can consider the situation that some nodes have a UDP flow and some have a TCP flow. If a node has both UDP flows and TCP flows, in a saturated situation, the aggregate traffic behaves similar to the UDP traffic.

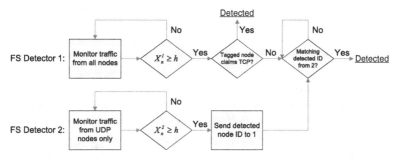

Fig. 2.15 Detection with TCP/UDP hybrid traffic

it claims to use TCP, we decide that it is a smart misbehaving node actually using UDP (case 3 mentioned above); if it claims to use UDP, we turn to listen to the decision from FS detector 2 (to avoid false positive in case 1). FS Detector 2 starts simultaneously with detector 1 but monitors only the traffic from the UDP nodes. When detector 2 identifies misbehavior from a UDP node, it sends the detected node ID to detector 1. If the detected node ID from detector 2 matches that alarmed by detector 1, the dual-detector system will then determine that the node is misbehaving (case 2). We run simulations to verify the performance of the dual-detector. For example, in a network of 10 nodes where 5 nodes use TCP and the other 5 nodes use UDP, the average false positive rate over a normal UDP node is 0.0047. Also, for the moderate misbehavior of $CW_{min} = 16$, the average detection delay of a misbehaving node is 18.1953 when it lies to be a TCP node. The detection delay increases to 36.4728 (for confirmed detection in both detectors) when the misbehaving node is honest with its UDP behavior, which is similar to that in the all UDP case listed in Table 2.1.

2.6 Summary

In this chapter, we develop a novel fair share (FS) detector for real-time backoff mis-behavior detection in IEEE 802.11^{TM} based wireless networks. Also, we develop a Markov chain based model to theoretically analyze the performance of the detector. While most existing work for backoff misbehavior detection depends on heuristic parameter configuration and experimental performance evaluation, we are able to use our model for a quantitative study to achieve guaranteed detection performance in terms of average false positive rate, average detection delay and missed detection ratio. Moreover, we present simulation results that confirm the accuracy of our theocratical analysis and demonstrate the robustness of the FS detector.

References

1. G. Bianchi, "Performance Analysis of the IEEE 802.11 Distributed Coordination Function," *IEEE Journal on Selected Areas of Communication*, vol. 18, no. 3, pp. 535–547, Mar. 2000.
2. B. Brodsky and B. Darkhovsky, *Nonparametric Methods in Change-point Problems*, Kluwer Academic Publisher, 1993.
3. C. E. Koksal, H. Kassab and H. Balakrishnan, "An Analysis of Short-Term Fairness in Wireless Media Access Protocols," in *Proc. ACM SIGMETRICS*, 2000.
4. J. R. Morris, *Markov Chains*, Cambridge Univ. Press, 1997.
5. J. Tang, Y. Cheng, and W. Zhuang, "An Analytical Approach to Real-Time Misbehavior Detection in IEEE 802.11 Based Wireless Networks," in *Proc. IEEE INFOCOM*, 2011.
6. The Network Simulator - ns-2, [Online.] Available: http://www.isi.edu/nsnam/ns/.
7. A. Toledo and X. Wang, "Robust Detection of Selfish Misbehavior in Wireless Networks," in *IEEE J. Sel. Areas Commun.*, vol. 25, no. 6, pp. 1124–1134, Aug. 2007.

Chapter 3
Adaptive Misbehavior Detection in IEEE 802.11™ Based on Markov Decision Process

Abstract To achieve better detection performance, we enhance the FS detector from Chap. 2 to develop an adaptive detector with the Markov decision process (MDP). In particular, we adaptively make decisions on how aggressively the detector value should be updated in each step. Then based on a reward function defined by us, we are able to determine an optimal decision policy to maximize the overall system benefit through a linear programming formulation. The optimal policy also indicates the operation of the adaptive detector, which yields better performance in both false positive rate and detection delay. Both theoretical analysis and simulation results are provided to demonstrate the performance of the adaptive detector.

3.1 Adaptive Detector Design

The FS detector (2.2) can be conveniently enhanced to incorporate adaptive operations. The basic idea is that at a certain state, if the tagged node is inferred to be misbehaving, the detector can increase its value more aggressively for a shorter detection delay; otherwise, the detector can decrease more aggressively to mitigate false positives. Let X_n denote the state of the detector at time n (i.e., the state after processing n observation samples). We use u_n to denote the adaptive actions associated with the state X_n. In this book, u_n can take the value of a positive integer, a negative integer, or 0. How to properly determine the action for each state will be discussed in the following section. With the adaptive actions, the behavior of the adaptive detector can be mathematically described as

$$X_{n+1} = (X_n + (NI_n + u_n - 1))^+$$
$$X_0 = 0. \qquad (3.1)$$

Also, the adaptive detector has the same decision rule as (2.3). And similarly, the detector value will be reset to 0 once a detection event happens.

J. Tang and Y. Cheng, *Intrusion Detection for IP-Based Multimedia Communications over Wireless Networks*, SpringerBriefs in Computer Science, DOI 10.1007/978-1-4614-8996-2_3, © The Author(s) 2013

3.2 Markov Decision Process Based Modeling

Consider the sequence of the adaptive detector value $\{X_n\}$ as a discrete random process, which takes values from a finite set $\mathscr{A} = \{0, 1, 2, \ldots, h\}$. The process is said to be in state i at time n if $X_n = i$. We can see that both the FS detector (2.2) and the adaptive detector (3.1) have the Markov property, that is, given the current state X_n, the next state X_{n+1} is independent of previous states. We have applied Markov chain analysis to study the FS detector in Chap. 2. We here apply the Markov decision process to study the adaptive detector.

Given an action $u_n = u$ associated with the state X_n, the transition probability can be expressed as

$$P_{ij}(u) = P\{X_{n+1} = j | X_n = i, u_n = u\}. \tag{3.2}$$

An MDP is a 4-tuple: $(\mathscr{A}, \mathscr{U}, P_{ij}(u), R_{ij}(u))$.

- $\mathscr{A}$ is a finite set of states. In this chapter, the detector value X_n is defined as the state and takes values from the set $\mathscr{A} = \{0, 1, 2, \ldots, h\}$.
- $\mathscr{U}$ is a finite set of actions. The action is the value of the adjustment u_n chosen at the state X_n. Given positive integer u_{max}, we consider the action set consisting of the integers in the range $[-u_{max}, u_{max}]$. How to select the action u_n according to the state X_n is called a *policy*.
- $P_{ij}(u)$ is the transition probability given that action u is taken at state X_n.
- $R_{ij}(u)$ is the reward received with the transition from state i to state j under the action u.

In the following, we will discuss how to determine the transition probability $P_{ij}(u)$, design the reward function $R_{ij}(u)$, as well as decide the optimal policy.

3.2.1 Transition Probability

Our observation measure is still I_n, the indicator of whether the current successful transmission over the network belongs to the tagged node. Thus, I_n's probability distribution can similarly be determined as in (2.1).

Using the distribution of I_n, and also taking into account the value of the detection threshold h and the action u, we can then calculate the transition probability $P_{ij}(u)$. Based on the operation of the adaptive detector, the calculation of $P_{ij}(u)$ is divided into four distinct cases.

Case 1   consists of $P_{ij}(u)$ for $i \in [0, h-1]$ and $j \in \{0, h\}$, with values

$$P_{ij}(u) = \begin{cases} 1 & \text{if } j = 0 \text{ and } i+N+u-1 \leq 0, \\ P\{I_n = 0\} & \text{if } j = 0, \ i+u-1 \leq 0 \text{ and} \\ & \quad i+N+u-1 > 0, \\ P\{I_n = 1\} & \text{if } j = h, \ i+u-1 < h \text{ and} \\ & \quad i+N+u-1 \geq h, \\ 1 & \text{if } j = h \text{ and } i+u-1 \geq h, \\ 0 & \text{otherwise.} \end{cases} \tag{3.3}$$

This case is related to transitions from any state other than state h to state 0 or h. According to the state transition equation (3.1), the detector variable X_n can only jump from the current state i to two other states, i.e., a larger value $i+N+u-1$ when $I_n = 1$ and a smaller value $i+u-1$ when $I_n = 0$. Thus, if the larger value $i+N+u-1 \leq 0$, X_n will for sure jump to 0; if the smaller value $i+u-1 \geq h$, X_n will for sure jump to h. Note that the state h in fact incorporates all possible states $X_n \geq h$, as the detector will raise an alarm when the state hits h. Further, if $i+u-1 \leq 0$ and $i+N+u-1 > 0$, X_n can only jump from i to 0 when $I_n = 0$. Also, if $i+N+u-1 \geq h$ and $i+u-1 < h$, X_n can only jump from i to h when $I_n = 1$.

Case 2   consists of $P_{ij}(u)$ for $i \in [0, h-1]$ and $j \in [1, h-1]$, with values

$$P_{ij}(u) = \begin{cases} P\{I_n = 0\} & \text{if } j = i+u-1 \text{ and} \\ & \quad i+u-1 > 0, \\ P\{I_n = 1\} & \text{if } j = i+N+u-1 \text{ and} \\ & \quad i+N+u-1 < h, \\ 0 & \text{otherwise.} \end{cases} \tag{3.4}$$

This case is related to the typical behavior of the detector, describing the transitions from any state other than state h to any state other than 0 and h. The state can transit to state $i+u-1$ when $I_n = 0$ or to state $i+N+u-1$ when $I_n = 1$, according to the state transition equation (3.1).

Case 3   consists of $P_{ij}(u)$ for $i = h$ and $j \in \{0, h\}$, with values

$$P_{hj}(u) = \begin{cases} 1 & \text{if } j = 0 \text{ and } N+u-1 \leq 0, \\ P\{I_n = 0\} & \text{if } j = 0, \ u-1 \leq 0 \text{ and} \\ & \quad N+u-1 > 0, \\ P\{I_n = 1\} & \text{if } j = h, \ u-1 < h \text{ and} \\ & \quad N+u-1 \geq h, \\ 1 & \text{if } j = h \text{ and } u-1 \geq h, \\ 0 & \text{otherwise.} \end{cases} \tag{3.5}$$

Case 4   consists of $P_{ij}(u)$ for $i = h$ and $j \in [1, h-1]$, with values

$$P_{hj}(u) = \begin{cases} P\{I_n = 0\} \text{ if } j = u - 1 \text{ and} \\ \qquad\quad u - 1 > 0, \\ P\{I_n = 1\} \text{ if } j = N + u - 1 \text{ and} \\ \qquad\quad N + u - 1 < h, \\ 0 \qquad\qquad \text{otherwise.} \end{cases} \qquad (3.6)$$

Cases 3 and 4 are related to transitions from state h. Such transitions are singled out as state h is special. According to the detector operation, we reset the detector value X_n to 0 right after it reaches h. This transition from h to 0 is not triggered by any successful transmission over the network and the two values happen in the same sampling period. This state has the value of h when it is entered and the value of 0 when it is left. Therefore, the transition probabilities as shown in Cases 3 and 4 are in fact equivalent to those associated with state 0 contained in Cases 1 and 2, respectively. Note that the policy at state h will also be the same as that at state 0. The reason for which we still maintain the state h is that it is the state to trigger the alarm of misbehavior detected.

Note that the Markovian model defined by (3.3)–(3.6) is an improvement over that defined by (2.5)–(2.7), as it more accurately models the transition from h to 0.

3.2.2 Reward Function

With an MDP formulation, the optimal policy at each state will be solved from an optimization problem that maximizes a reward function. Thus, the reward function needs to be properly designed. It should have the property that a positive reward will be collected for a right action that can improve the performance, while a negative penalty will be given to an improper action that degrades the performance. Regarding our problem of misbehavior detection, the evaluation of an action will depend on the analysis of the node behavior. We use $P\{M|X_n = i\}$ to denote the probability that the tagged node is misbehaving, given the current detector state $X_n = i$. We propose to define a reward function for an action u in state i as

$$R\{i, u\} = -(1 - P\{M|X_n = i\})u + P\{M|X_n = i\}u. \qquad (3.7)$$

We can see that the reward function will encourage choosing a positive value for the action u when the tagged node is misbehaving, represented by the positive reward $P\{M|X_n = i\}u$, where the positive u represents an aggressive increase towards fast detection. On the other hand, the reward function stimulates the selection of a negative value for u when the tagged node is a normal one, represented by the positive reward $-(1 - P\{M|X_n = i\})u$, where the reward is for an aggressive decrease to mitigate the false positives. The reward function is represented as a probabilistic average, considering the randomness in practice. It is not a trivial issue to calculate the probability $P\{M|X_n = i\}$ though.

We consider a general multiple misbehaving nodes scenario for the analysis. Specifically, we assume that each node in a network of N nodes could be a misbehaving one independently with a probability of Q, i.e., $P\{M\} = Q$. With good majority, normally $Q < 50\%$. We further assume that when a node is misbehaving, it randomly chooses the value of its minimum contention window CW_{min} from a set $\mathcal{W}$, with a uniform probability of $\frac{1}{|\mathcal{W}|}$, where $|\mathcal{W}|$ is the cardinality of the set $\mathcal{W}$. The CW_{min} defined in the IEEE DCF standard is 32, and all the elements $\mathcal{W}$ are less than 32. According to the contention windows selected, the nodes in the network can be divided into S classes. All the nodes in the same class use the same CW_{min}. Specifically, we denote that a class i node uses $CW_{min} = W_i$ with $W_i \in \{\mathcal{W}, 32\}$. If each class has N_i (≥ 0) nodes, we have $\sum_{i=1}^{S} N_i = N$.

With our model, we are to calculate the probability $P\{X_n = i\}$, which is the steady state probability that the detector X_n stays in state i when monitoring a node, averaging all possible misbehaving cases in the network. We will also calculate the probability $P\{X_n = i|M\}$, which is the steady state probability that the detector X_n stays in state i, given that a misbehaving node is monitored. Note that the two probabilities will be calculated with $u = 0$, which represents the performance if no extra adaptive action is taken and serves as the standpoint for the optimal policy design. The probabilities will allow us to determine the reward function through calculating the probability

$$P\{M|X_n = i\} = \frac{P\{X_n = i|M\}P\{M\}}{P\{X_n = i\}}. \tag{3.8}$$

According to the classic modeling approach for the 802.11^{TM} DCF [1], we consider that each node independently accesses an idle channel for transmission. Let p_t^i denote the probability that a class i node transmits at a random time slot and p_c^i denote the collision probability of a given transmission from a class i node. We further let m denote the maximum backoff stage. Given the number of nodes in each class $N_1, \cdots, N_S$ and the CW_{min} taken by each class $W_1, \cdots, W_S$, according to [1], we have the following equations:

$$\begin{cases} p_t^1 = \dfrac{2(1 - 2p_c^1)}{(1 - 2p_c^1)(W_1 + 1) + p_c^1 W_1(1 - (2p_c^1)^m)} \\[2ex] \quad\vdots \\[1ex] p_t^S = \dfrac{2(1 - 2p_c^S)}{(1 - 2p_c^S)(W_S + 1) + p_c^S W_S(1 - (2p_c^S)^m)} \\[2ex] p_c^1 = 1 - (1 - p_t^1)^{N_1 - 1} \displaystyle\prod_{i=2}^{S}(1 - p_t^i)^{N_i} \\[2ex] \quad\vdots \\[1ex] p_c^S = 1 - (1 - p_t^S)^{N_S - 1} \displaystyle\prod_{i=1}^{S-1}(1 - p_t^i)^{N_i} \end{cases} \tag{3.9}$$

from which the parameters p_t^i, p_c^i of each class i node can be solved.

Note that a node can get a successful transmission under the circumstance that there is no collision while the node transmits. Thus from the solutions of (3.9), we can obtain the probability that a node gets a successful transmission at a random time slot:

$$p_s^1 = p_t^1(1 - p_c^1), \tag{3.10}$$

$$\vdots$$

$$p_s^S = p_t^S(1 - p_c^S). \tag{3.11}$$

Suppose that the tagged node belongs to class v. We can then calculate the probability q_s^v that a successful transmission over the network is from the tagged node with $CW_{min} = W_v$ as

$$q_s^v = \frac{p_s^v}{\sum_{i=1}^S N_i p_s^i}. \tag{3.12}$$

Note that q_s^v is a conditional probability given a specific configuration of CW_{min} and number of nodes in each class, i.e.,

$$q_s^v(\cdot) = f(W_1, \cdots, W_S, N_1, \cdots, N_S). \tag{3.13}$$

We can further analyze other conditional and unconditional successful transmission probability resorting to our misbehaving model. Recall that in our model every node misbehaves with a probability Q and the node who misbehaves randomly chooses its CW_{min} from the set $\mathcal{W}$. Let $P(W_i)$ denote the probability that a node sets its CW_{min} as W_i ($\in \mathcal{W}$). We then have

$$P(W_i) = \begin{cases} \dfrac{Q}{|\mathcal{W}|} & \text{if} \quad W_i \in \mathcal{W} \\ 1 - Q & \text{if} \quad W_i = 32 \end{cases}. \tag{3.14}$$

Let $\bar{q}_s(W_v)$ denote the conditional probability that a successful transmission is from the tagged node, given that the tagged node has a CW_{min} size of W_v. Let $\bar{q}_s$ denote the unconditional probability that a successful transmission is from the tagged node, averaging over all the possible network scenarios. We can see that

$$\bar{q}_s(W_v) =$$

$$\sum_{\substack{W_i \in \mathcal{W}: \\ i \neq v \text{ and} \\ W_i \neq W_j \text{ for } i \neq j}} \sum_{\substack{(N_1, \cdots, N_S): \\ N_v \geq 1 \text{ and} \\ \Sigma_i N_i = N}} q_s^v(\cdot) P(W_v)^{(N_v - 1)} \prod_{i=1, i \neq v}^S P(W_i)^{N_i} \tag{3.15}$$

and further

$$\bar{q}_s = \frac{Q}{|\mathcal{W}|} \sum_{W_v \in \mathcal{W}} \bar{q}_s(W_v) + (1 - Q)\bar{q}_s(32). \tag{3.16}$$

In (3.15) and (3.16), we consider the different combinations of the CW_{min} taken by each class and the number of nodes in each class.

Recall that (2.1) is the probability distribution of our observation measure I_n for a tagged node. Applying $\bar{q}_s$ to (2.1), we can obtain the probability distribution of I_n. We can then calculate $P\{X_n = i\}$ with $u = 0$ as mentioned above. By using the I_n distribution in (3.3)–(3.6), we can compute the transition probabilities $P_{ij}(0)$. Let $(\pi_0, \ldots, \pi_h)$ denote the steady state probabilities when $u = 0$, which can be solved from the equations

$$\pi_j = \sum_{i=0}^{h} \pi_i P_{ij}(0), \quad j \in \{0, \ldots, h\}, \tag{3.17}$$

$$\sum_{j=0}^{h} \pi_j = 1. \tag{3.18}$$

We can then get $P\{X_n = i\} = \pi_i$ from (3.17) and (3.18).

Similarly, if we apply $\bar{q}_s(W_v)$ obtained in (3.15)–(2.1) and further compute the transition probabilities and then solve the equations (3.17) and (3.18) based on such transition probabilities, we will obtain the steady state probability $P\{X_n = i | CW_{min} = W_v\}$, which is the distribution of the detector state given that the tagged node takes a CW_{min} of W_v. Then as the tagged node may chooses its CW_{min} uniformly from $\mathcal{W}$ when it is misbehaving, we can calculate $P\{X_n = i | M\}$ as

$$P\{X_n = i | M\} = \frac{P\{X_n = i, M\}}{P\{M\}}$$

$$= \frac{1}{|\mathcal{W}|} \sum_{W_v \in \mathcal{W}} P\{X_n = i | CW_{min} = W_v\}. \tag{3.19}$$

With the probabilities $P\{X_n = i | M\}$ and $P\{X_n = i\}$, we can calculate the probability $P\{M | X_n = i\}$ according to (3.8) and then obtain the reward function by (3.7).

3.2.3 Optimization Problem Formulation

Our goal is to determine the optimal policy, i.e., how to choose the action at a certain state, to achieve the maximum benefit based on the reward function developed above. In particular, we will find the steady-state probability π_{iu} of being in state i and choose action u when the optimal policy is used. Hence, given that the state space is determined by a certain detection threshold h, i.e., $i, j \in [0, h]$, the problem can be formulated as

$$\max \sum_i \sum_u \pi_{iu} R(i, u)$$

$$\text{subject to } \sum_i \sum_u \pi_{iu} = 1,$$

$$\sum_u \pi_{ju} = \sum_i \sum_u \pi_{iu} P_{ij}(u) \quad \text{for all } j,$$

$$\pi_{ju} \geq 0 \quad \text{for all } i, u. \tag{3.20}$$

This is in fact a linear programming problem. Solving (3.20) using the simplex method, we can obtain the set of π_{iu}^* maximizing the overall benefit, which also indicates the optimal policy of how to choose an action u at state i. We also find that for each i, π_{iu}^* is zero for all but one value of u, which is due to the property of the linear programming problem (Sect. 4.10 of [15]). Thus there is only one action, i.e., u^*, to be taken for a state in the optimal policy, which indicates how our adaptive detector will operate under a certain detection threshold h. Then we compare the total reward of various detection thresholds and select an h with the largest reward to be our detection threshold. As an example, for a network with $N = 8$ nodes, each of which tends to misbehave with a probability of $Q = 0.25$, we determine a detection threshold of $h = 20$ which achieves the largest reward, and also obtain the optimal configuration u^* associated with the h.

3.3 Theoretical Performance Analysis

In this section, we conduct theoretical performance analysis of the adaptive detector whose operation is characterized by the optimal policy obtained from last section, in terms of three fundamental metrics: *average false positive rate*, *average detection delay*, and *missed detection ratio* under a detection delay bound. In fact, with the optimal policy for each state obtained, the MDP will reduce to a Markov chain. When we analyze this Markov chain in a specific scenario, we can numerically evaluate the performance of the adaptive detector in that case. In this section, we will also compare the performance of the adaptive detector to that of the FS detector.

3.3.1 Average False Positive Rate

The average false positive rate P_{fp} is the rate that the detector value X_n hits state h given the fact that no node in the network is misbehaving. This rate is equal to the steady-state probability that the Markov chain describing the adaptive detector stays at h in the normal condition. In the normal condition with a fair share of the channel access, we have $q_s^v = \frac{1}{N}$ for a tagged node. We can calculate the distribution of I_n according to (2.1). And further, using the optimal u for each state obtained from (3.20), we calculate the transition probability matrix $\mathbf{P}^*$ of the Markov chain characterizing the behavior of the adaptive detector under the normal traffic condition (normal Markov chain), according to (3.3)–(3.6). Then similarly through (2.8)–(2.10), we can get the average false positive rate P_{fp} as in Sect. 2.4.1.

As an example, for a network with $N = 8$ nodes, each of which tends to misbehave with a probability of $Q = 0.25$, we determine a detection threshold of $h = 20$ which achieves the largest reward, and also obtain the optimal configuration of u associated with the h. Then based on the Markov chain model obtained from the optimization, we compute P_{fp} as 0.0013.

3.3.2 Average Detection Delay

The average detection delay denoted as $E[T_D]$ is the average number of samples observed from the moment that the tagged node starts to misbehave until the misbehavior is detected. With the Markov chain under the abnormal condition (abnormal Markov chain), $E[T_D]$ can be computed as the expected number of transitions required for the state variable to hit state h, starting from the moment when the misbehavior starts. As in Sect. 2.4.2, to obtain $E[T_D]$, we need to find the transition probabilities of the abnormal Markov chain and determine the initial state of the detector when the misbehavior starts.

Given the CW_{min} of a misbehaving node $\hat{v}$, we can calculate the probability that a successful transmission is from $\hat{v}$, $\hat{q}_s^v$, through (3.9)–(3.12). Using $\hat{q}_s^v$ in (2.1), we can obtain the probability distribution of $\hat{I}_n$ for the misbehaving node. And further using the optimal u for each state obtained from (3.20), the transition probability matrix $\hat{P}^*$ for the abnormal Markov chain can be obtained according to (3.3)–(3.6). Next based on the normal Markov chain from Sect. 3.3.1, the initial state probabilities can be determined through (2.15). Then subsequently with the transition probabilities of the abnormal Markov chain and initial state probabilities both in place, the average detection delay $E[T_D]$ can be calculated in a similar way as in Sect. 2.4.2.3 through (2.16) and (2.17).

Table 3.1 Average detection delay of the adaptive detector with $P_{fp} = 0.0013$, $h = 20$

	$W_2 = 2$	$W_2 = 4$	$W_2 = 8$	$W_2 = 16$
$W_1 = 2$	12.8151	4.321	4.096	4.0168
$W_1 = 4$	$1.98 * 10^4$	16.2949	7.3324	6.1042
$W_1 = 8$	$7.11 * 10^5$	270.0457	31.1042	17.4963
$W_1 = 16$	$1.59 * 10^7$	$5.61 * 10^3$	307.9787	111.3373

As an example, we consider a network also with $N = 8$ nodes; among them there are two nodes, denoted as node 1 and node 2, misbehaving. We use the same optimal parameter configuration as in the example in Sect. 3.3.1. We treat node 1 as the tagged node and compute $E[T_D]$ for detecting the node. Varying the CW_{min} of both of the two misbehaving nodes, the delays for detecting node 1 are shown in Table 3.1. Here W_1 denotes the CW_{min} of node 1 and W_2 denotes the CW_{min} of node 2. From the table, as expected, we can see that more intense misbehavior of node 1, i.e., a smaller W_1, leads to a shorter detection delay for node 1; however, more intense misbehavior of node 2 leads to a longer detection delay for node 1. Note that when $W_1 = W_2$, the two misbehaving nodes are "fairly" competing with each other to transmit. But once W_2 becomes larger than W_1, the detection delay for node 1 incurs a drastic decline, as node 1 starts to gain advantages over node 2 and gets more opportunities to transmit. Moreover, we see that for the cases where $W_1 > W_2$, the detection delays for node 1 become very large since node 2 is gaining advantages over node 1. Such long detection delays do not make much practical sense. We will discuss our method for dealing with the multiple misbehaving nodes scenarios in Sect. 3.3.4. And later in our simulation results, we will demonstrate the effectiveness of the method.

We then compare the detection performance of the adaptive detector to the performance of the FS detector with the static operation of $u = 0$ in each state, to examine whether the adaptive detector has indeed achieved better performance. Here we still consider the setting where there are 2 misbehaving nodes in a network of $N = 8$ nodes.

Table 3.2 Average detection delay of the FS detector with $P_{fp} = 0.0013$, $h = 70$

	$W_2 = 2$	$W_2 = 4$	$W_2 = 8$	$W_2 = 16$
$W_1 = 2$	22.4527	9.9944	9.5412	9.3783
$W_1 = 4$	$2.90 * 10^7$	26.1878	15.2037	13.2251
$W_1 = 8$	$5.67 * 10^{12}$	174.044	38.721	27.3762
$W_1 = 16$		$3.18 * 10^5$	198.6957	85.3163

Table 3.2 shows the detection delays of the FS detector for detecting node 1, giving the same false positive constraint of $P_{fp} = 0.0013$. To achieve this false positive rate, the FS detector needs to set its detection threshold as $h = 70$. Comparing the results in Table 3.2 to Table 3.1, we can see that for most cases of $W_1 \leq W_2$, the adaptive detector can achieve much quicker detection delay than the FS detector. The reason for this better performance is that when the detector is in smaller states, the optimal policy mostly instructs it to choose a negative u; whereas in larger states, the optimal policy mostly instructs the detector to choose a positive u. As a result, it is more difficult for the detector to increase its value in the normal condition since initially the detector value X_n is small and there are relatively few transmissions from the tagged node to increase the detector's value. However, when the tagged node starts to misbehave, consecutive transmissions from the node trigger X_n to become large, and the optimal policy at this time make the increase even greater, resulting in quicker detection delay. Overall, the adaptive detector is able to achieve better performance in both false positive and detection delay. Note that we do not show the result for the case of $W_1 = 16$ and $W_2 = 2$, as the number is too large to be considered as practically detectable.

Table 3.3 shows the detection delays of the FS detector for detecting node 1, giving the same detection threshold of $h = 20$. With this threshold, the false positive rate for the FS detector is $P_{fp} = 0.0138$, which is about one order larger than the false positive rate achieved for the adaptive detector of $P_{fp} = 0.0013$. Comparing the results in Table 3.3 to Table 3.1, we can see that with the same detection threshold,

Table 3.3 Average detection delay of the FS detector with $P_{fp} = 0.0138, h = 20$

	$W_2 = 2$	$W_2 = 4$	$W_2 = 8$	$W_2 = 16$
$W_1 = 2$	6.1162	2.725	2.6108	2.5712
$W_1 = 4$	803.9305	7.0898	4.1654	3.6144
$W_1 = 8$	$1.74 * 10^4$	33.6002	10.2318	7.3958
$W_1 = 16$	$2.34 * 10^5$	284.2008	36.3525	20.3005

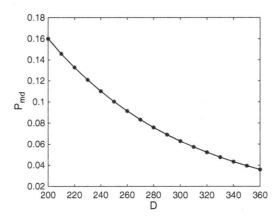

Fig. 3.1 Missed detection ratio for 1 node with $CW_{min} = 16$

the FS detector does have some advantage over the adaptive detector in detection delay for most cases of $W_1 \leq W_2$. However, this advantage in detection delay of the FS detector is not quite comparable to the advantage of the adaptive detector in false positive rate. This further justifies the strength of the adaptive detector.

3.3.3 Missed Detection Ratio

As in Sect. 2.4.3, the missed detection ratio P_{md} is examined under a given detection delay constraint D, since the adaptive detector also has the property of "non-stop until detection". The detection event happens only when X_n hits state h. Thus the missed detection ratio P_{md} under the delay constraint D is the summation of the probabilities of X_n staying at a state other than h at time D. Then based on the normal Markov chain from Sect. 3.3.1 and the abnormal Markov chain from Sect. 3.3.2, P_{md} can similarly be determined through (2.18)–(2.22) as in Sect. 2.4.3.

Figure 3.1 demonstrates the missed detection ratio P_{md} of detecting 1 of 2 misbehaving nodes both with $CW_{min} = 16$, in a network of 8 nodes as obtained from our analysis. As shown in the figure, to detect such moderate misbehavior, we can set $D = 330$ to obtain a P_{md} less than 0.05.

3.3.4 Multiple Misbehaving Nodes Scenario

In cases where multiple misbehaving nodes exist in a wireless network, our detector can detect the misbehaving nodes one by one. We monitor the transmissions from every node using a separate detector. As the node with the most intense misbehavior has the most opportunities to transmit, we will be able to quickly detect it with our detector. After that, we will discard packets from that node and continue the detection among the leftover nodes. Clearly, the node with the second most intense misbehavior will be detected at this time. We will continue such detection iteratively to identify every misbehaving node. The effectiveness of this approach will be demonstrated in Sect. 3.4.1.

What needs to be noted here is that we are not assuming eliminating the wireless channel access of those nodes with more intense misbehavior after they are detected, as practically it is not easy to achieve such elimination, and the detected misbehaving nodes can continue transmitting packets to keep on impacting contentions in the network. The specifics of how to eliminate misbehaving nodes, however, are out of the scope of this book. In our scheme, whether to eliminate the nodes from the network or not is not a problem for continuing detecting other nodes with less intense misbehavior. The detector can just drop the packets from those nodes already detected so that the detector is shielded from the impact of those nodes. This is an special advantage of our detector over existing misbehavior detectors [19, 20] in detecting multiple misbehaving nodes. The observation measures of those detectors have to include information from the whole network, and thus transmissions from any node in the network will always have impact on the detection of a tagged misbehaving node. The "shielding" option is not available to those detection methods.

3.4 Simulation Results

We first establish an 802.11^{TM} DCF based wireless network consisting of 8 competing nodes ($N = 8$) and an AP through ns-2 simulation [18]. The network works under the saturated condition and every node transmits packets over the User Datagram Protocol (UDP) towards the AP. The AP also acts as the detection agent which monitors the transmissions from every competing node with a separate detector. The nodes are located close enough to sense the transmissions from each other and thus avoid the hidden terminal problem. There are 2 misbehaving nodes, marked as node 1 and node 2, among the 8 competing nodes, the same as in the theoretical analysis. The 2 misbehaving nodes can manipulate their minimum contention window to a value from $\{2, 4, 8, 16, 32\}$.

Through simulation, we obtain average detection delays of the adaptive detector, and the results for detecting node 1 are shown in Table 3.4. Note that we do not include the cases of $W_1 > W_2$ in this table, as most of them are not practically detectable in the first place. We will discuss this issue in next subsection.

Table 3.4 Average detection delay of the adaptive detector for $W_1 \leq W_2$ with $h = 20$

	$W_2 = 2$	$W_2 = 4$	$W_2 = 8$	$W_2 = 16$
$W_1 = 2$	8.3378	5.518	5.5385	5.6583
$W_1 = 4$		9.9695	5.6956	5.5825
$W_1 = 8$			20.2828	12.655
$W_1 = 16$				79.2084

Given the same detection threshold of $h = 20$, we compare the simulation results in Table 3.4 to the analytical results in Table 3.1. Even though the two sets of results are close, generally the simulation results are smaller than the analytical results. The reason is that our analysis is based on the assumption of independent channel access; however, in a practical 802.11™ network, a node that has just accomplished a successful transmission will have advantages in grabbing the channel for next transmission in a short period [4], which implies correlations among the channel accesses. As a result, a selfish misbehaving node can obtain even more channel accesses in addition to those resulting from the misbehavior. Therefore, the average detection delays obtained from simulations are shorter. However, the analytical results still provide a conservative estimation for the detector's performance, which is meaningful for us to guide the detection system configuration.

Table 3.5 Average detection delay of the adaptive detector for $W_1 > W_2$ with $h = 20$

	$W_2 = 2$	$W_2 = 4$	$W_2 = 8$	$W_2 = 16$
$W_1 = 2$				
$W_1 = 4$	38.3886			
$W_1 = 8$	70.3503	66.9261		
$W_1 = 16$	243.7508	197.8860	131.2686	

3.4.1 Multiple Misbehaving Nodes Scenario

In the cases of $W_1 > W_2$, the misbehaving node 1 is practically not detectable simultaneously with node 2 as its channel accesses are much less than node 2. However, as node 2 can be detected quickly in these cases, after the detection of node 2, we can simply discard the transmissions from node 2 and only continue our detection among the remaining 7 nodes to detect node 1. As we monitor the transmissions from every competing node with a separate detector, this approach is very easy to implement. Table 3.5 shows the results obtained from the approach. The detection delays are the original delays to detect node 2 plus the additional delays to detect node 1 after transmissions from node 2 are discarded. As we can see, the results are significantly smaller than the analytical results from Table 3.1, which makes the detection of multiple misbehaving nodes much more feasible.

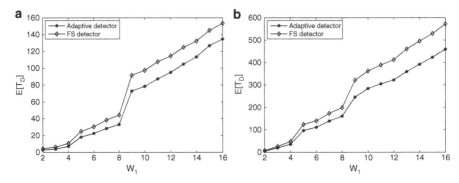

Fig. 3.2 Average detection delay in multiple misbehaving nodes scenarios. (**a**) $W_2 = 4$, $W_3 = 8$, $W_4 = 16$. (**b**) $W_2 = 2$, $W_3 = 4$, $W_4 = 8$

Then to be more general, we consider a network of 20 nodes with 4 nodes misbehaving, one of which is the tagged node. The tagged node varies its minimum contention window size W_1 from 2 to 16, while the other 3 misbehaving nodes set their windows as W_2, W_3 and W_4 respectively to represent a wide range of misbehavior. Figure 3.2a shows the average delays for detecting the tagged node when $W_2 = 4$, $W_3 = 8$ and $W_4 = 16$, representing a more moderate misbehaving environment; Fig. 3.2b shows the results when $W_2 = 2$, $W_3 = 4$ and $W_4 = 8$, representing a more intense misbehaving environment. The sudden increases of delays in the figures, e.g., when $W_1 = 5$ and 9 in Fig. 3.2a, are due to the reason that starting from these window values, there will be one more misbehaving node detected before the tagged node, which obviously increases the delay. Overall, we can see that the adaptive detector is faster than the FS detector in all the misbehaving cases. Also as expected, more intense misbehavior from other nodes makes it longer to detect the tagged node.

3.5 Summary

In this chapter, we develop an adaptive approach to address real-time selfish misbehavior detection in IEEE 802.11^{TM} based wireless networks. By enhancing the FS detector from Chap. 2, we design an adaptive detector where actions are added to control how the detector value is updated. Further, we model the adaptive detector using the Markov decision process (MDP), which enables us to find an optimal policy to determine the proper action to be taken in each state of the adaptive detector. The optimal policy characterizes the operation of the adaptive detector, which also enables us to theoretically analyze the detection performance. The simulation results confirm the accuracy of our analysis, and also demonstrate the detector's ability to address the cases of multiple misbehaving nodes.

3.6 Related Work of Selfish Misbehavior Detection in 802.11^{TM}

The problem of detecting backoff misbehavior over 802.11^{TM} based wireless networks has been studied in various scenarios and under several mathematical frameworks in the literature. In [7, 8], the authors present a modification to the 802.11^{TM} protocol to facilitate the misbehavior detection. In their scheme, the receiver assigns a backoff timer for the sender. If the number of idle slots between consecutive transmissions from the sender does not comply with the assigned backoff timer, the receiver will consider that the sender potentially deviates from the protocol and penalize the sender with a smaller backoff timer. Continuous deviations will result in that the receiver labels the sender as a selfish node. However, the scheme assumes a trustworthy receiver which performs the detection, which may not be the case in a dynamic network environment. Modifications to the 802.11^{TM} protocol and reliance on the receiver are the main limitations of the work.

Another approach to deal with the backoff misbehavior is to develop protocols based on the game-theoretic techniques, by imposing adequate costs on network operations [3, 5, 6]. The goal is to encourage all the nodes to reach a Nash equilibrium. As a result, a malicious node is not able to gain an unfair share compared to well-behaved nodes and thus discouraged from the misbehavior. However, this category of approaches assume that all the nodes are willing to deviate from the protocol when necessary, and performance of the network may converge to a suboptimal operation point. Moreover, modifications to the standard protocols are also required.

A heuristic sequence of conditions are proposed in [12, 13] to test multiple misbehavior options in the 802.11^{TM} protocol based on simple numerical comparisons. The detection algorithm estimates the average values of the option parameters and raises alarms when the cumulative effect of the misbehavior exceeds a threshold. This approach, named DOMINO, preserves its advantage of simplicity and easiness of implementation, and still demonstrates its efficiency when dealing with a wide range of 802.11^{TM} MAC misbehavior. However, the heuristic nature of the approach limits its applications to specific protocols.

The sequential probability ratio test (SPRT) method is used in [10, 11, 14] to detect the 802.11^{TM} backoff misbehavior. The detection decision is made when a random walk of the likelihood ratio of observations (given two hypotheses) rises to be larger than an upper threshold. The main advantage of SPRT is that it can reach decision very fast, given the complete knowledge of both normal behavior and backoff misbehavior strategy [9]. However, in a realistic setting, the strategy of malicious nodes is hard to know in advance; such an issue imposes the major inherent limitation of the SPRT method. Further, the existing work normally assumes that the backoff timer of each node is observable, which is again hard to achieve in practice because the transmission attempts involved in a collision are impossible to be distinguished. In our design, we monitor the successful transmission of the tagged node as the observation measurement.

The authors in [19, 20] utilize the Kolmogorov–Smirnov (K–S) significance test for backoff misbehavior detection. This test is able to make the decision by measuring the distribution of the idle time between consecutive successful transmissions from a tagged node and comparing it to the normal backoff behavior. The detection method in [19, 20] requires estimation of the collision probability of a packet transmitted. However, an inaccurate simplification there is to consider that packets from the misbehaving node and those from the normal nodes have the same collision probability. Such inaccuracy impacts both the performance of false positive rate and detection delay, as demonstrated in Sect. 2.5.3. Furthermore, as a batch test method, the K–S statistic has its own drawback. Fixed-size data samples are needed to perform the test each time, which makes real-time detection difficult.

In our preliminary work [16, 17], we adopt the non-parametric CUSUM test [2] for the backoff misbehavior detection, which has the advantages of both real-time detection and no requirement of a priori knowledge of the misbehavior strategy. The detector in [17] directly counts the number of successful transmissions from a tagged node within an observation window[1] to get a sample. Although such a sampling method is easy for implementation, the observation window needs to linearly increase with the number of nodes in the network to fairly count transmissions from each node, which as a result will increase the detection delay. In this book, our FS detector and adaptive detector take every successful transmission over the network as a sample to trigger its state change. Such a sampling method is independent of the network size and turns out to result in improved performance in both false positive rate and detection delay.

A common research issue among most of the existing schemes for misbehavior detection is their dependency on heuristic parameter configuration and experimental performance evaluation, which largely limits the flexibility and robustness of these schemes. To address this issue, in this book, we develop a Markov chain based analytical model to theoretically analyze the detection performance and quantitatively configure the system parameters. We also resort to the Markov decision process technique to provide theoretical guidance on the adaptive design for further detection performance improvement.

References

1. G. Bianchi, "Performance Analysis of the IEEE 802.11 Distributed Coordination Function," *IEEE Journal on Selected Areas of Communication*, vol. 18, no. 3, pp. 535–547, Mar. 2000.
2. B. Brodsky and B. Darkhovsky, *Nonparametric Methods in Change-point Problems*, Kluwer Academic Publisher, 1993.
3. M. Cagalj, S. Ganeriwal, I. Aad and J. Hubaux, "On Cheating in CSMA/CA Ad Hoc Networks," Technical Report IC/2004/24, EPFL-DI-ICA, 2004.
4. C. E. Koksal, H. Kassab and H. Balakrishnan, "An Analysis of Short-Term Fairness in Wireless Media Access Protocols," in *Proc. ACM SIGMETRICS*, 2000.
5. J. Konorski, "Protection of Fairness for Multimedia Traffic Streams in a Non-cooperative Wireless LAN Setting," in *PROMS*, vol. 2213 of LNCS. Springer, 2001.

---

[1] An observation window is defined as a certain number of consecutive successful transmissions over the whole network [17].

6. J. Konorski, "Multiple Access in Ad-Hoc Wireless LANs with Noncooperative Stations," in *NETWORKING*, vol. 2345 of LNCS. Springer, 2002.

7. P. Kyasanur and N. Vaidya, "Detection and Handling of MAC Layer Misbehavior in Wireless Networks," in *Proc. IEEE DSN*, 2003, pp. 173–182.

8. P. Kyasanur and N. Vaidya, "Selfish MAC Layer Misbehavior in Wireless Networks," in *IEEE Trans. Mobile Comput.*, vol. 4, no. 5, pp. 502–516, 2005.

9. H. V. Poor and O. Hadjiliadis, *Quickest Detection*, first edition, Cambridge, 2008.

10. S. Radosavac, J. S. Baras and I. Koutsopoulos, "A Framework for MAC Protocol Misbehavior Detection in Wireless Networks," in *Proc. ACM Workshop on Wireless Security*, 2005, pp. 33–42.

11. S. Radosavac, G. Moustakides, J. Baras and I. Koutsopoulos, "An Analytic Framework for Modeling and Detecting Access Layer Misbehavior in Wireless Networks," in *ACM Trans. Information and Systems Security*, vol. 11, no. 4, article no. 19, Jul. 2008.

12. M. Raya, J. Hubaux and I. Aad, "DOMINO: A System to Detect Greedy Behavior in IEEE 802.11 Hotspots," in *Proc. ACM MobiSys*, 2004.

13. M. Raya, I, Aad, J. Hubaux and A. El Fawal, "DOMINO: Detecting MAC Layer Greedy Behavior in IEEE 802.11 Hotspots," in *IEEE Trans. Mobile Computing*, vol. 5, no. 12, pp. 1691–1705, Dec. 2006.

14. Y. Rong, S. Lee and H. Choi, "Detecting Stations Cheating on Backoff Rules in 802.11 Networks using Sequential Analysis," in *Proc. IEEE INFOCOM*, 2006, pp. 1–13.

15. S. M. Ross, *Introduction to Probability Models*. Elsevier Academic Press, 9th edn., 2007.

16. J. Tang, Y. Cheng, Y. Hao and C. Zhou, "Real-Time Detection of Selfish Behavior in IEEE 802.11 Wireless Networks," in *Proc. IEEE VTC-Fall*, 2010.

17. J. Tang, Y. Cheng, and W. Zhuang, "An Analytical Approach to Real-Time Misbehavior Detection in IEEE 802.11 Based Wireless Networks," in *Proc. IEEE INFOCOM*, 2011.

18. The Network Simulator - ns-2, [Online.] Available: http://www.isi.edu/nsnam/ns/.

19. A. Toledo and X. Wang, "A Robust Kolmogorov-Smirnov Detector for Misbehavior IEEE 802.11 DCF," in *Proc. IEEE ICC*, 2007, pp. 1564–1569.

20. A. Toledo and X. Wang, "Robust Detection of Selfish Misbehavior in Wireless Networks," in *IEEE J. Sel. Areas Commun.*, vol. 25, no. 6, pp. 1124–1134, Aug. 2007.

Chapter 4
SIP Flooding Attack Detection

Abstract In this chapter and the following chapter, we address the SIP layer attack detection. In this chapter we focus on the well-known flooding attack and develop an online scheme to detect and subsequently prevent the attack, by integrating a novel three-dimensional sketch design with the Hellinger distance detection technique.

4.1 SIP Flooding Attack

4.1.1 Multimedia Communications with SIP

IP-Based multimedia communications utilizes SIP [5] as the application-layer signaling protocol to establish, manage and terminate communication sessions. At the transport layer, SIP normally favors the user datagram protocol (UDP) over the transmission control protocol (TCP) due to the simplicity of UDP and the connection-oriented nature of SIP itself. There are three basic components in a SIP environment, which are user agent client (UAC), user agent server (UAS) and SIP proxy server. These components are identified using the SIP address, which has a similar form to an email address, typically containing a username and a host name, e.g., "sip:alice@iit.edu". Messages are exchanged between these components to perform ordinary SIP operations.

The SIP messages used to establish and terminate sessions are basically INVITE, 200 OK, ACK and BYE. They are also called the SIP methods or attributes. A UAC initiates a SIP session by sending out an INVITE. Intermediate proxies look over the destination SIP address in the message and forward it to the destined UAS who will respond with a 200 OK. An ACK message then finishes the three-way handshake to establish the session and media will go directly between the UAC and the UAS. When the session is finished, it will be terminated by a BYE message from either of the calling parties.

J. Tang and Y. Cheng, *Intrusion Detection for IP-Based Multimedia*
Communications over Wireless Networks, SpringerBriefs in Computer Science,
DOI 10.1007/978-1-4614-8996-2_4, © The Author(s) 2013

4.1.2 Flooding Attack

SIP is vulnerable to network anomalies such as the flooding attacks. These attacks can be easily mounted by utilizing various SIP traffic generators openly available on the Internet, e.g., SIPp [7]. The victim SIP proxy servers can be overwhelmed or even crushed by a large number of SIP messages within a short period of time.

SIP utilizes multiple methods/attributes to manage sessions. This provides possibilities for the attackers to take advantage of the vulnerabilities of these attributes to launch different forms of SIP flooding attacks. We describe some of these attacks below. We see that a general detection/prevention system is desired to defend these attacks.

4.1.2.1 INVITE Flooding

In this attack, thousands of INVITE messages are generated and transmitted to the victims which can barely support all of them. Moreover, being a transactional protocol, SIP may require the intermediate proxy servers to hold resource for each IN-VITE message when they are expecting the associated 200 OK. Thus the resources of these victim proxy servers could be exhausted almost in real time if the attack rate is high enough.

4.1.2.2 BYE Flooding

The BYE message is used to terminate SIP sessions. Therefore it can be utilized by the attackers to bring down ongoing multimedia communications sessions. More severely, the attackers can just launch a brute force BYE flooding attack to prematurely tear down most ongoing sessions in a network without the knowledge of the SIP addresses of the legitimate users. Such flooding attacks will cause call drops over a big range of users immediately.

4.1.2.3 Multi-attribute Flooding

Intelligent attackers can launch different forms of SIP flooding attacks together to the victim proxy servers in a distributed manner. In this case, not only will the resources of the proxy servers be exhausted, but all the ongoing sessions may also be torn down instantly at the same time, which makes the multi-attribute flooding attacks devastating to the multimedia communications service. Moreover, the attacks flood the four SIP attributes simultaneously and thus do not change the relative proportions of the attributes. Therefore the existing SIP flooding detection solution [6] based on observing significant deviations in such proportions will become ineffective against the multi-attribute flooding attacks.

4.2 Basic Techniques

Our flooding detection and prevention system monitors the SIP messages arriving at a proxy server. We implement it in a firewall module, which can be deployed without modifying the proxy server. The system operation is based on two techniques, sketch and Hellinger distance.

4.2.1 Sketch

The sketch data structure is a probabilistic data summarization technique. It builds compact and fixed-size summaries of high dimensional data streams through random aggregation, by applying a hash function [8] to the data. Specifically, we consider that each data item consists of a key k_i and its associated value v_i [3], represented as $a_i = (k_i, v_i)$, for constructing a sketch. Data items whose keys are hashed to the same value will be put in the same entry in sketch and their values will be added up to obtain the value of that entry. In our scheme, we use the SIP address as the key, and the value associated with each key is set as 1 indicating one SIP attribute generated from that address.

Using sketch makes our scheme scalable. No matter how many users exist in a network, sketch is able to derive a fixed-size traffic summary. More importantly, sketch allows us to construct a probability distribution based on the sketch entries, with no need to investigate the correlation among different SIP attributes as described in [6].

4.2.2 Hellinger Distance

The Hellinger distance (HD) is used to measure the distance between two probability distributions [9]. To compute HD, suppose that we have two histogram distributions on the same sample space, namely, $P = (p_1, p_2, \cdots, p_n)$ and $Q = (q_1, q_2, \cdots, q_n)$. The HD between the two distributions is defined as follow

$$H^2(P,Q) = \frac{1}{2} \sum_{i=1}^{n} (\sqrt{p_i} - \sqrt{q_i})^2. \tag{4.1}$$

It is not difficult to see that the HD will be up to 1 if the two probability distributions are totally different and down to 0 if they are identical. This property provides a good approach to quantify the similarity of two data sets in either normal or anomalous situations. We aim to build an anomaly detection system which needs a statistical model to represent the normal traffic condition and raises alarms when abnormal variations are observed. The property of HD makes it well suited to this role. A low

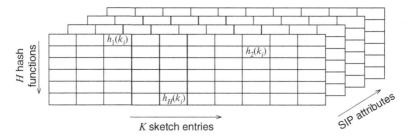

Fig. 4.1 Illustration of a three-dimensional sketch design

HD value implies that there is no significant deviation in the current traffic observations and a high HD is a strong indication that anomalies have happened.

4.3 Detection and Prevention Scheme Design

In this section we describe our scheme to detect and prevent the SIP flooding attacks. The scheme is based on integrating the two techniques introduced in Sect. 4.2, sketch and Hellinger distance.

4.3.1 Three-Dimensional Design

The SIP flooding attack can bear different forms and thus induce changes in multiple SIP attributes. We must be able to isolate the changes across the attributes, then discriminate the diverse attack forms and cope with the multi-attribute attack.

Figure 4.1 gives an illustration of our three-dimensional sketch design. The sketch comprises multiple two-dimensional attribute hash-tables, each of which is built for a SIP attribute. We build four such tables for the four SIP attributes investigated. An attribute hash-table consists of H element hash-rows, each of which is associated with a different hash function and has K entries. We construct the hash functions using independent random seeds [8], and therefore they are independent from each other. The hash functions are kept secret because the seeds are not known to others. The three-dimensional sketch design allows us to separately summarize each of the SIP attributes. In the following, we first discuss how to calculate an HD based on each hash-row, and then describe the operation in the context of three-dimensional sketch.

We divide time into discrete intervals and each interval is of a constant length d. The messages associated with a certain SIP attribute under consideration is indexed as a data stream. The data stream then passes through two periods: a training period and a test period. The training period contains T consecutive time intervals and the test period is the $(T+1)$th interval. We build two sketches, one for the training period and the other for the test period. The SIP address of each message is used as the key for the data to be put into the sketch. Such two sketches can generate two probability distributions for HD analysis.

Based on the training set, we obtain a sketch data distribution P. Suppose that the values of the K entries are $n_1, n_2, \cdots, n_K$, and we denote $N = \sum_{i=1}^{K} n_i$. Then we define the distribution P as

$$P = (\frac{n_1}{N}, \frac{n_2}{N}, \cdots, \frac{n_K}{N}). \tag{4.2}$$

Similarly, we obtain a distribution Q based on the sketch for the test period. Suppose that the values of the K entries of the test sketch are $m_1, m_2, \cdots, m_K$, with $M = \sum_{i=1}^{K} m_i$. We can have the distribution Q as

$$Q = (\frac{m_1}{M}, \frac{m_2}{M}, \cdots, \frac{m_K}{M}). \tag{4.3}$$

The Hellinger distance of the above two distributions is then calculated as

$$H^2(P,Q) = \frac{1}{2} \sum_{i=1}^{K} (\sqrt{\frac{n_i}{N}} - \sqrt{\frac{m_i}{M}})^2. \tag{4.4}$$

We monitor the data stream by tracing the HD. Assume that there is no attack in the first training set, which initially represents the normal condition. To calculate the HD, we obtain the "test" distribution Q from the current time interval and the "training" distribution P from the immediately preceding T time intervals. We continue this operation and move the test and training periods forward respectively at each time interval, as long as the HD is smaller than a threshold. Such a sliding window mechanism better estimates the pattern of the data stream than directly analyzing two consecutive individual time intervals. It can well reflect the dynamics of the evolving traffic and smooth sudden fluctuations in normal traffic.

All the H hash-rows in an attribute hash-table independently monitor the data stream associated with a certain SIP attribute, following the same operation as described above. Similarly, in the three-dimensional sketch, the four attribute hash-tables investigate the four SIP attributes separately and are prepared for the attack detection.

4.3.2 Threshold Under Attack

4.3.2.1 Detection Threshold

As we want to utilize HD to model the traffic behavior along time, a detection threshold is needed to reflect the normal condition and be the actual indicator of anomalies. Since normal traffic behaviors also fluctuate over time and the distribution obtained based on sketch may even not be stationary, the HD in the normal condition will be non-zero and may dynamically change. In order to properly model the behavior, we adopt the exponential weighted moving average (EWMA) method [2] in our scheme to compute a dynamic threshold.

Let h_n denote the value of HD in the current time interval n. To smooth its fluctuation, we calculate an estimation average, H_n, of h_n as

$$H_n = (1 - \alpha) \cdot H_{n-1} + \alpha \cdot h_n. \tag{4.5}$$

Next, to have an estimate of how much H_n deviates from h_n, we compute the current mean deviation S_n as

$$S_n = (1 - \beta) \cdot S_{n-1} + \beta \cdot |H_n - h_n|. \tag{4.6}$$

Then given values of H_n and S_n, we derive the estimated threshold H_{n+1}^{Thre} following

$$H_{n+1}^{Thre} = \lambda \cdot H_n + \mu \cdot S_n, \tag{4.7}$$

where λ and μ are multiplication factors used to set a safe margin for the threshold. Due to the ability of HD to accurately monitor the difference between two probability distributions, proper values of these two parameters may greatly reduce false alarms. The parameters α, β, λ and μ are all tunable parameters in the model. We set the initial values of them according to previous research [6] and tune them in our experiments to achieve desirable detection accuracy.

4.3.2.2 Estimation Freeze Mechanism

When the HD obtained from a certain element hash-row exceeds the threshold, an attack detection is registered. After this, if we continue the update according to (4.5)–(4.7), the threshold will be polluted by the attack as the attacking traffic will be taken into account in estimating the threshold. To avoid this from happening, we freeze the threshold and keep it as a constant as long as the HD is above it. Also, to prevent the attacking traffic from entering the training set and thus keep the HD high only during attacks, we modify the sliding window mechanism. As shown in Fig. 4.2, after an attack detection is registered at the $(i+1)$th time interval d_{i+1}, we freeze the current training set and only let the test set proceed to the next time interval. This "one freezing one proceeding" action only ends when the HD goes below the threshold and the normal sliding window is then resumed. Overall, the above operations are illustrated in Algorithm 1, termed by us as the "estimation freeze mechanism". As a side benefit of the mechanism, we can determine the attack duration D because the HD is above the threshold all through the attack and immediately comes down right afterwards.

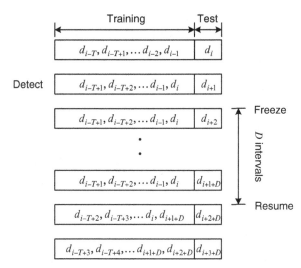

Fig. 4.2 Sliding window in estimation freeze mechanism

Algorithm 1: Estimation Freeze Mechanism

Input: SIP attribute stream
Output: Duration of the anomaly D
$D = 0$;
d = time interval length;
anomaly starting time $t_1 = 0$;
anomaly ending time $t_2 = 0$;
if *HD surpasses threshold* **then**
    t_1 = time of *HD* surpassing threshold;
    $t_2 = t_1$;
    freeze training set;
    freeze threshold;
    while *HD > threshold* **do**
        proceed test set;
        calculate *HD* between test set and freezed training set;
        $t_2 = t_2 + d$;
    end
    $D = t_2 - t_1$;
    else
        proceed training set;
        proceed test set;
        update threshold;
    end
end
return D;

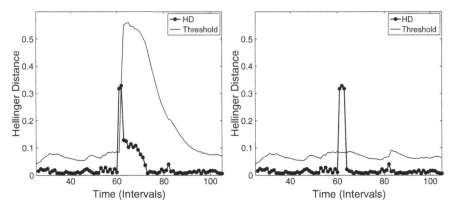

Fig. 4.3 Comparison of thresholds under attack

We illustrate a comparison between two thresholds under attack in the same traffic condition in Fig. 4.3. The left one is estimated directly from HD without our estimation freeze mechanism whereas the right one is obtained using the mechanism. We see that without freezing the threshold goes all the way up with HD when the attack is detected. It is even much higher than HD after the detection and can not reflect the normal traffic condition. Obviously such a threshold mechanism loses track of the attack after the initial detection. On the contrary, using our estimation freeze mechanism, the threshold remains low and HD keeps high after the attack is detected. Together they also explicitly determine the duration of the attack. This provides a very clear indication of the entire attack.

4.3.3 Attack Detection

As described above, to actually detect possible attacks, the HD associated with a certain hash-row will be computed between the sketch data distribution constructed from the test set and that constructed from the training set. In an ideal case, assuming that the normal probability distribution could be accurately measured from the training set, the threshold for detection can be set as 0. We have the following theorem.

Theorem 4.1: A flooding attack over a SIP attribute can be detected with high probability by computing the HD between sketch data distributions, assuming that the normal probability distribution could be accurately measured from the training set.

Proof: Consider an element hash-row in the attribute hash-table under investigation. Suppose that the hash-row has K entries. The total volume of normal traffic in the test set is M, which is distributed into the K entries according to $M = \sum_{i=1}^{K} m_i$, with m_i denoting the volume counted by the ith entry. Assume that there is a flooding attack with total volume M' added over the normal traffic in the test set, which is distributed to K' ($\leq K$) entries according to $M' = \sum_{i=1}^{K'} m_i'$. Let p_i denote the probability mass of entry i, and $p_i = \frac{m_i}{M}$ in the normal situation. Assume that the entry is contaminated by the attacking traffic, the probability mass will then be $p'_i = \frac{m_i + m'_i}{M + M'}$. Assume that the training set can accurately monitor the normal probability distribution and the test set is consistent with such a distribution. The performance of the HD-based detection is then determined by the relation between p_i and p'_i as

$$\left| p_i - p'_i \right| = \left| \frac{m_i}{M} - \frac{m_i + m'_i}{M + M'} \right| = \left| \frac{m_i M' - m'_i M}{M(M + M')} \right|$$

$$= \left| \frac{\frac{m_i}{M} - \frac{m'_i}{M'}}{1 + M/M'} \right|. \tag{4.8}$$

Given a threshold of 0, the attacker needs to set the distribution of the flooding traffic exactly as the normal distribution to avoid being detected. A significant benefit of utilizing the sketch data distribution is that the hash functions used by the detection system will be kept secret to users. Therefore, it is hard for the attackers to estimate the normal sketch data distribution even if they can monitor the raw user data distribution. Furthermore, the detection system can dynamically change the sketch hash functions for a higher level of security. If the attacker attempts to guess the normal sketch data distribution $\frac{m_i}{M}$, the probability of guessing the correct value will be low, because the value of $\frac{m_i}{M}$ in a given entry can be considered as a *continuous* random variable. In other words, our detection system can detect the attack with high probability. ∎

Theorem 4.1 demonstrates the ideal performance under accurate distribution modeling. Practically, since random aggregation of sketch brings information loss and normal traffic itself is dynamic, the normal probability distribution may change over time. Thus we can not monitor it that ideally in practice and detection accuracy may be impacted due to approximation. However, the analysis shows us that attacks will indeed disturb the probability distribution obtained from the test set and as a result cause HD to rise.

In an attribute hash-table, each element hash-row registers attacks independently when its associated HD exceeds the detection threshold. Random aggregation in sketch brings information loss in each hash-row. Thus in one detection circle certain rows may register attacks whereas others may not. However, if most rows agree on an attack, it is highly likely that the attack actually occurs. Correspondingly, if only a small portion of rows find one attack, we can probably consider it as a false alarm. Thus, to increase detection confidence and assure high accuracy, we apply a voting procedure: if at least z percent of the H rows in an attribute hash-table register attacks, a flooding attack alarm is finally raised.

4.3.4 Attack Prevention

After detecting the flooding attack, the next step is to identify the offending SIP messages and selectively discard them to prevent the attack from reaching the proxy servers and causing damages. In order to achieve this, we first identify the anomalous sketch entries that contain the offending messages in each row. Assuming that the normal probability distribution could be accurately measured from the training set, we have the following theorem.

Theorem 4.2: In a flooding attack context, when the HD-based detection indicates an attack, there must exist entries in a sketch hash-row for the test set which has a larger probability mass than that in the corresponding entry for the training set, and such entries are definitely associated with certain offending SIP messages.

Proof: In the normal situation, we assume that the normal probability distribution could be accurately measured from the training set and the test set is consistent with the distribution. Thus, we have $\frac{m_i}{M} = \frac{n_i}{N}$. In the context under attack, the probability mass deviation in an entry i is

$$p'_i - \frac{n_i}{N} = p'_i - \frac{m_i}{M} = \frac{\frac{m'_i}{M'} - \frac{m_i}{M}}{1 + M/M'} \tag{4.9}$$

according to (4.8). When the HD detection indicates an attack, there must exist entries where $p'_i \neq \frac{n_i}{N}$. Moreover, in such entries, we must have $p'_i > \frac{n_i}{N}$ for some of them and $p'_i < \frac{n_i}{N}$ for others; otherwise the condition that $\sum_{i=1}^{K} p'_i = 1$ could not be maintained. In those entries with $p'_i > \frac{n_i}{N}$, the item associated with offending messages $\frac{m'_i}{M'}$ must exist. However, the entries with $p'_i < \frac{n_i}{N}$ may not include offending messages. The reason is that the attacking traffic might only occupy a subset of the entries in a hash-row, i.e., $K' < K$. In the leftover $K - K'$ entries, $m'_i = 0$ and offending messages are not included. ∎

According to Theorem 4.2, we mark entries whose probability increases as possible anomalous entries. Suppose that we have p_i as the probability mass of the ith entry in one row from the training sketch set and q_i as the probability mass of the same entry from the test set. Then, if the condition

$$\sqrt{p_i} - \sqrt{q_i} < 0 \tag{4.10}$$

satisfies, we mark this ith entry as a suspicious entry. We use square roots of p_i and q_i since we have already obtained the value of every $\sqrt{p_i} - \sqrt{q_i}$ when we calculate HD. Therefore this operation would not incur much more computational cost to our scheme.

Let U_j denote the set of SIP messages that are mapped to the suspicious entries of the jth row in an attribute hash-table. We then tag these messages in U_j as offending message candidates. Certainly there will be normal SIP messages among these candidates because sketch hashes multiple users to one entry. However, since each row in a table independently performs random aggregation, offending messages and certain normal messages which are hashed to the same entry in one row are not likely to be hashed to one entry in other rows. Thus, we identify the offending message set U over all the H rows in a table through

$$U = \bigcap_{j=1}^{H} U_j. \tag{4.11}$$

This intersection of candidates filters out normal messages in the suspicious entries. As a result, the set U are finally believed to just include the offending SIP messages.

Once the offending messages are identified, they will be immediately discarded and only normal SIP messages can go through. This ensures that the proxy servers will only serve normal messages, and also effectively prevent the attacks from reaching the proxy servers and subsequently causing damages.

4.4 Performance Evaluation

We evaluate the performance of the proposed SIP flooding detection and prevention scheme in this section. As VoIP is one of the most prominent and representative application of IP-based multimedia communications, in our simulation, SIP signaling traces are generated based on signaling in a VoIP network and analyzed through Matlab. We demonstrate the accuracy of the scheme on both attack detection and prevention. The analysis focuses on the INVITE flooding case first since other SIP attributes can be addressed in a similar way. We also investigate the advantage of our scheme over the detection scheme in [6] where the effectiveness of the scheme [6] can be severely affected by the combination effect of dynamic normal traffic arrival and call holding time. Then we extend our discussion to the cases of distributed denial of service (DDoS) attack and multi-attribute attack.

4.4.1 Normal Traffic Behavior

In the normal condition, the average call generating rate is uniformly distributed from 25 per second to 75 per second with a mean of 50 per second. The senders of the messages are chosen from 100,000 users. The numbers of messages from each sender are long tail distributed which more likely resembles the real life situation. The reason for this is that in an operational VoIP network, users do not fairly make phone calls with the same frequency. Most calls come from a relatively small number of heavy users, whereas the rest lighter users who are the majority do not make phone calls that frequently. Therefore we model the user SIP addresses using the Pareto distribution [4]. Also, to properly model the BYE messages, we set normal call holding times to I, where I either follows a log-normal distribution to reflect the long tail characteristic of real VoIP call holding times [1] or equals to a constant 60 s according to [6]. We will discuss the effect of I later.

We parse INVITE messages from the trace data. As in [6], to achieve higher detection accuracy and lower computational cost, we set the length of a time interval d to 10 s. Also, as longer training set better captures the pattern of the traffic whereas

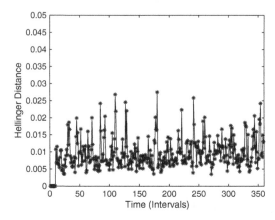

Fig. 4.4 Hellinger distances under normal traffic

shorter training set responds quicker to change, in order to find a good balance between them, the number of time intervals in a training set T is set to 10.

We build two sketches for both the training set and the test set and calculate the Hellinger distances between their related element hash-rows along time as described in Sect. 4.3. As shown in Fig. 4.4, the HDs are mostly distributed around 0.01 when we choose $K = 32$ and $H = 5$. These low HD values show the similarity of the training set and the test set when the traffic behaviors are normal.

4.4.2 Ineffectiveness of Rate-Based Approach

In the flooding attack experiment, we use the normal traffic described above as background and mix it with the flooding traffic from an attacking source. In Fig. 4.5, we show the dynamics of traffic rates when there are five attacks of 50 INVITEs per second from a single attacker mixed with the normal traffic. The durations of the attacks are all 30 s. We see that there is hardly a sign of abnormal behaviors in the figure since the normal traffic itself has fluctuation as well. Comparatively, we will see how our scheme responds to these attacks in next section.

4.4.3 Flooding Attack Detection and Prevention

4.4.3.1 Detection

We apply our scheme to detect the same five attacks of 50 INVITEs per second as described above. We set the initial values of the parameters in the scheme according to previous research [6] and empirically get their final values as $\alpha = 0.125$,

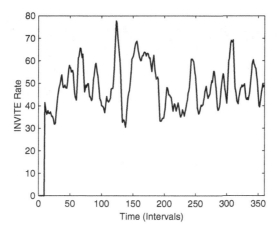

Fig. 4.5 Dynamic traffic rate

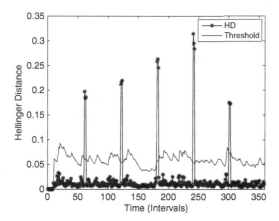

Fig. 4.6 Detection of flooding attacks

$\beta = 0.25$, $\lambda = 5$, $\mu = 1$ to achieve desirable detection accuracy. Figure 4.6 shows the dynamics of the HD obtained from a hash-row and the associated threshold. The five spikes clearly identify the five flooding attacks. Other rows may not have the same detection accuracy due to different aggregations of INVITE messages, but as we set $z = 80\%$, the voting procedure finds an agreement among the five rows and raises attack alarms accurately. Also in Fig. 4.6, due to the "estimation freeze mechanism" applied, we can see that the HD remains high and threshold keeps constant during attack. They together precisely determine the duration of an attack, which lasts for 3 time intervals, i.e., 30 s. Both the HD and the threshold evolve with the dynamics of the traffic and thus preserve the ability to detect attacks online. Whereas in [6], the threshold does not react accordingly under attack and remains low as if it is always estimated from normal traffic. Compared to our threshold mechanism, theirs is not able to accurately reflect the online traffic situation.

Table 4.1 Detection results

Flooding rate	Number of experiments	Detection probability
15	50	88 %
35	50	100 %
50	50	100 %
75	50	100 %
100	50	100 %
500	50	100 %

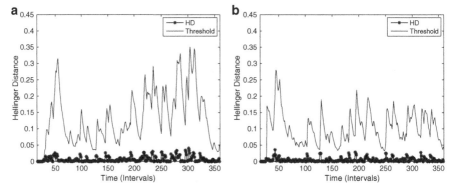

Fig. 4.7 Limitation of existing detection scheme [6]. (**a**) Dynamic traffic rate and constant I. (**b**) Dynamic traffic rate and lognormal I

We repeat the experiment for several times and change the attack rates accordingly. The flooding rates vary from 15 per second to 500 per second. Fifteen per second is lower than the minimal normal rate whereas 500 INVITE messages per second can quickly overwhelm a SIP proxy server. The purpose of choosing such a wide range is to see that despite effectively detecting high rate flooding, our scheme is even capable of identifying low rate attacks which can hide in the normal traffic and still preserves high accuracy. The durations of the attacks are all 30 s. The detection results are shown in Table 4.1. We can see that our scheme is able to detect the attacks with 100 % accuracy when the attack rate is as low as 35 per second. Also, even when the attack rate comes down to 15 per second, we are still able to detect 88 % of the attacks.

In [6], probability distributions are derived by monitoring the relative proportions of the four SIP attributes within the same period of time. However, as BYE comes after a relatively long lag, i.e., the call holding time I, compared to the other three attributes, its number within a certain period of time is correlated to the number of the other three attributes which arrived I seconds ago. Thus if the normal traffic arrival rate is dynamic, the probability distribution derived from the relative proportions of the four SIP attributes within the same period of time will certainly have great fluctuations and result in large deviation between the training set and the test set even under the normal condition. Figure 4.7a illustrates the HD and the associated threshold calculated based on the scheme in [6] from the same traffic condition where we obtain Fig. 4.6. The normal average call generating rate is uniformly distributed from 25 per second to 75 per second with a mean of 50 per second and there are five attacks of 50 INVITEs per second mixed with the normal traffic. The call holding time I is set to a constant $60\,s$ in this case. We see that the HD is relatively large even when traffic is normal before and between attacks, with a mean value of $5 * 10^{-3}$. Also, the five instances of attacks are not detected, as the attacks can not bring larger deviation compared to the normal traffic. We find that for the scheme in [6] to be more effective, the standard deviation of the normal traffic rate needs to be small. We have similar observations from Fig. 4.7b, where all the setting is the same as Fig. 4.7a except that I is set to follow a lognormal distribution.

Through investigation, we learn that dynamic traffic arrivals can severely affect the effectiveness of the scheme in [6] as BYE needs to arrive later due to call holding times. Comparatively, our scheme establishes probability distributions and detects attacks over each attribute independently, which eliminates the dependency on the correlation between different attributes. Call holding times do not affect our scheme and high detection accuracy is achieved even under dynamic traffic arrivals. Therefore, our detection scheme is more effective and flexible than the scheme in [6].

4.4.3.2 Prevention

For attack prevention, our scheme accurately identifies all the offending INVITE messages from the single attacker and can thus drop the messages to prevent the attacks from damaging the network. There is no missed identification in each of the attack occasions. There are two facts contributing to this high accuracy. First, all the offending messages are aggregated to just one suspicious entry in each of the element hash-rows. Second, the intersection of the five suspicious entries respectively from the five element hash-rows is enough to filter out all the involved normal messages and identify the offending ones.

4.4.4 DDoS Attack Detection and Prevention

4.4.4.1 Detection

In the case of the DDoS attack, numerous attackers initiate flooding to a SIP proxy server simultaneously. To test our scheme against such attacks, we launch five DDoS from 300 attackers through simulation. Figure 4.8 shows that the attacks cause obvious deviation in HD along time. The five spikes of HD in the figure clearly identify the attacks. We run the experiment multiple times varying attacker numbers and rates. The results show neither missed detection nor false alarm. The principle behind this high detection accuracy is that sketch randomly aggregates data in the H element hash-rows and deploys the same number of totally independent probability

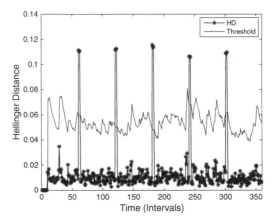

Fig. 4.8 Detection of DDoS attacks

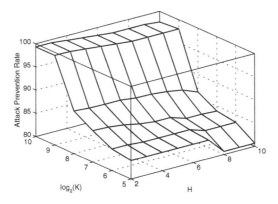

Fig. 4.9 DDoS prevention rate varying K and H

distributions. Also, the numbers of messages from each sender are long tail distributed. These make it impossible for the attackers to capture the pattern of every
sketch data distribution deployed from the normal traffic in a large VoIP network.

4.4.4.2 Prevention

For the following attacker prevention, numerous offending SIP messages are identified by our scheme. However, there are still some missed offending messages that
we are not able to identify. This should be due to the fact that the K used by us is
smaller than the attacker number. In this case, every entry in one sketch row contains
offending messages, thus by applying (4.10) we can only find part of them.

We investigate the missed identification problem in the DDoS prevention operation. We vary the values of K and H to find out how they affect the prevention rate. The results are illustrated in Fig. 4.9. Clearly, when K increases, missed

identification decreases accordingly. As K becomes larger than the attacker number 300, we achieve almost 100 % accuracy. Changing the value of H does not affect the result much. However, a larger H here leads to more missed identification since more rows tend to have less consensus. Also, a small H can cause more false positives. Thus we figure that an H of 5 is a good enough value.

4.4.5 Multi-attribute Attack

We generate distributed multi-attribute flooding attacks through simulation. There are ten attack occasions in this experiment. In each occasion, attackers send a large number of the four SIP attributes, namely INVITE, 200 OK, ACK and BYE simultaneously. Results of the experiment show that our scheme successfully identifies the ten attack occasions of each SIP attribute. We build three-dimensional sketch data sets to separately address each attacking attribute. Thus our scheme is able to naturally discriminate the different forms of SIP flooding no matter which attribute is being used to launch the attacks.

4.5 Summary

In this chapter, we develop an online SIP flooding detection and prevention scheme by integrating two techniques, sketch and Hellinger distance. Our three-dimensional sketch design is capable of summarizing each SIP attribute separately and deploying associated probability distributions. Based on these distributions, the Hellinger distance is utilized to monitor the normal traffic behaviors and detect attacks if any abnormal variations are observed. Also, the "estimation freeze mechanism" presented shows its ability to both maintain the information about normal behavior under attack and determine the durations of the flooding attacks. A voting procedure is applied to assure the detection accuracy. Moreover, we utilize the random aggregation property of sketch and the consensus between all the rows to selectively discard the offending SIP message and subsequently prevent the attack. As we establish probability distributions for each SIP attribute independently, our scheme is fully effective to the multi-attribute attack. Our experimental results show that the scheme developed preserves high accuracy on both attack detection and prevention.

References

1. F. Gustafson and M. Lindahl, "Evaluation of statistical distributions for VoIP traffic modelling," University Essay from University West, Department of Economics and IT, 2009.
2. J. Kurose and K. Ross, *Computer Networking: A Top-Down Approach (4th ed.)*, Addison Wiley, 2007.

3. S. Muthukrishnan, "Data Streams: Algorithms and Applications," in *Proc. the Fourteenth Annual ACM-SIAM Symposium on Discrete Algorithms*, 2003.
4. W. Reed, "The Pareto, Zipf and Other Power Laws," *Economics Letters*, vol. 74, no. 1, pp. 15–19, Dec. 2001.
5. J. Rosenberg, H. Schulzrinne and G. Camarillo, "SIP: Session Initiation Protocol," IETF RFC 3261, Jun. 2002.
6. H. Sengar, H. Wang, D. Wijesekera and S. Jajodia, "Detecting VoIP Floods Using the Hellinger Distance," *IEEE Trans. Parallel Distrib. Syst.*, vol. 19, no. 6, pp. 794–805, Jun. 2008.
7. SIPp, [Online.] Available: `http://sipp.sourceforge.net/`.
8. M. Thorup and Y. Zhang, "Tabulation Based 4-Universal Hashing with Applications to Second Moment Estimation," in *Proc. the Fifteenth Annual ACM-SIAM Symposium on Discrete Algorithms*, 2004.
9. G. Yang and L. Le Cam, *Asymptotics in Statistics: Some Basic Concepts*, second edition, Wiley, Mar. 2006.

Chapter 5
SIP Stealthy Attack Detection and Resource-Drained Malformed Message Attack Detection

Abstract In this chapter, we first address the stealthy attack, where intelligent attackers can afford a long time to attack the system, and only incur minor changes to the system within each sampling period. To identify such attacks in the early stage for timely responses, we propose a detection scheme based on the signal processing technique wavelet, which is able to quickly expose the changes induced by the attacks. Then, we address the malformed message attack identified by us, which manipulates both the "Session-Expires" header in the SIP message and openness of wireless protocols to severely drain the network resources. We develop a detection method based on the Anderson–Darling test to deal with such attacks.

5.1 Stealthy Attack Detection

5.1.1 Stealthy Attack

The stealthy SIP flooding attack is launched by intelligent and patient attackers who start their attacks with no rush using a low initial rate r_0. And then they will periodically increase the current attack rate r_n in a slow pace following

$$r_n = r_{n-1} + \Delta r \tag{5.1}$$

where Δr is the increment of the attack rate at each burst period T. The stealthy attack does not cause sudden directly observable changes in traffic. Also, attackers can manipulate a smaller Δr and a longer T to further hide the attack behavior. The attack can bring damages to the network in a long time scale even though initially it may seem harmless.

J. Tang and Y. Cheng, *Intrusion Detection for IP-Based Multimedia Communications over Wireless Networks*, SpringerBriefs in Computer Science, DOI 10.1007/978-1-4614-8996-2_5, © The Author(s) 2013

5.1.2 Detection Scheme Design

Our scheme to detect the stealthy SIP flooding attack mainly consists of two steps, namely, the sketch traffic summarization and the wavelet-based attack detection. Sketch provides fixed-size manageable raw traffic signals no matter how many users exist in the network. And wavelet analysis extracts information from the raw signals to quickly expose changes brought by the stealthy attack. The two steps are performed after every sampling period Δt to identify whether there is attack existing in the period.

5.1.2.1 Sketch Traffic Summarization

As described in Sect. 4.2, the sketch data structure is a probabilistic data summarization technique. It randomly aggregates high dimensional data streams into a fixed-size summary with smaller dimensions through the use of hash operations.

Each input data item $a_i = (k_i, v_i)$ to sketch consists of a key k_i and its associated value v_i. When a new data item a_i arrives, its key is first hashed through $h(k_i)$, which becomes the index of a_i in the sketch table. The associated value v_i will then be added to the entry with the index $h(k_i)$ of the table. In our scheme, we use SIP address as the key and the number of INVITE messages from the address as the associated value.

The hash function $h(x)$ randomly aggregates INVITE messages from multiple SIP addresses into one entry. Therefore, suppose that the original message stream has a high dimension of X, applying the hash function can reduce its dimension to a smaller fixed number K. We use one sketch table to summarize the traffic stream from one sampling period Δt and the resultant "sketched" data will be used in the following wavelet analysis.

The random aggregation of sketch comes with a cost of information loss due to the use of hash operations. A remedy for this can be using multiple sketch tables each of which is associated with an independent hash function similarly as in Sect. 4.3. Wavelet analysis could then be applied separately to every sketch table and the final detection result would be based on the consensus among them. However, as we will see, just one hash table is already able to achieve high detection accuracy. Thus it is not necessary to use the multiple table approach.

The sketch traffic summarization is crucial for our detection. Even though the number of users in a network is dynamic, sketch can ensure that raw traffic signals from all sampling periods are of the same length, which provides great convenience for the wavelet-based detection. We will also see the impact of sketch on detection in our experimental results.

5.1.2.2 Wavelet-Based Detection

Suppose that the raw traffic signal obtained from sketch of a sampling period is S, which has K elements. We use wavelet analysis to decompose S into an approximation signal A and a detail signal D, and monitor the percentage of energy corresponding to D for attack detection.

As S has limited number of elements, we perform wavelet analysis through the Discrete Wavelet Transform (DWT). In particular, we choose one of the most commonly used DWT, the Daub4 [6] transform due to its simplicity and also its effectiveness in identifying the stealthy flooding attacks. The Daub4 transform has two sets of coefficients, the scaling coefficients $\{\alpha_1, \alpha_2, \alpha_3, \alpha_4\}$ and the wavelets coefficients $\{\beta_1, \beta_2, \beta_3, \beta_4\}$. The coefficients are pre-defined constants [6] and satisfy the relationship $\beta_k = (-1)^{k-1}\alpha_{4-(k-1)}$. Using the scaling coefficients, the Daub4 scaling signals V can be expressed as a $\frac{K}{2} \times K$ matrix

$$V = \begin{pmatrix} \alpha_1 & \alpha_2 & \alpha_3 & \alpha_4 & 0 & 0 & \dots & 0 & 0 \\ 0 & 0 & \alpha_1 & \alpha_2 & \alpha_3 & \alpha_4 & \dots & 0 & 0 \\ \vdots & \vdots & \vdots & \vdots & \vdots & \vdots & \ddots & \vdots & \vdots \\ \alpha_3 & \alpha_4 & 0 & 0 & 0 & 0 & \dots & \alpha_1 & \alpha_2 \end{pmatrix}.$$

Similarly, using the wavelets coefficients $\{\beta_1, \beta_2, \beta_3, \beta_4\}$, we can express the Daub4 wavelets also as a $\frac{K}{2} \times K$ matrix W.

Applying V and W to transform S, we obtain the trend signal and the fluctuation signal

$$a_i = \sum_{j=1}^{K} S_j V_{ij} \quad \text{for } i \in \{1, 2, \dots, \frac{K}{2}\} \tag{5.2}$$

$$d_i = \sum_{j=1}^{K} S_j W_{ij} \quad \text{for } i \in \{1, 2, \dots, \frac{K}{2}\}. \tag{5.3}$$

Next, using the trend signal and the fluctuation signal as coefficients, we calculate the approximation signal A and the detail signal D through

$$A_j = \sum_{i=1}^{\frac{K}{2}} a_i V_{ij} \quad \text{for } j \in \{1, 2, \dots, K\} \tag{5.4}$$

$$D_j = \sum_{i=1}^{\frac{K}{2}} d_i W_{ij} \quad \text{for } j \in \{1, 2, \dots, K\}. \tag{5.5}$$

Then our detector, the percentage of energy corresponding to the detail signal D is

$$P^d = \frac{\sum_{j=1}^{K}(D_j)^2}{\sum_{j=1}^{K}(A_j)^2 + \sum_{j=1}^{K}(D_j)^2}. \tag{5.6}$$

P^d keeps low under the normal traffic condition. However, when the stealthy flooding attack starts, P^d will immediately have a sharp increase even though there has not been directly observable changes in traffic. Thus monitoring P^d enables us to quickly identify whether there is attack existing in the network.

A threshold is needed to define the detection rule for each sampling period. To accurately keep track of the normal traffic condition, we use a dynamic threshold h based on the Exponential Weighted Moving Average (EWMA) method. Suppose that p_{n-1}^d is the smoothened P^d and v_{n-1} is the mean deviation of the previous sampling period computed through EWMA. We have the threshold h_n of the current sampling period as

$$h_n = \lambda \cdot p_{n-1}^d + \mu \cdot v_{n-1} \tag{5.7}$$

where λ and μ are multiplication factors used to set a safe margin for the threshold and thus avoid false alarms. Also, to maintain the information about normal traffic behavior under attacks, we freeze h as a constant once it is exceeded by P^d and only restart to update h again after P^d comes below. As a side benefit, this "threshold freezing scheme" also helps us trace the durations of attacks, because only during attacks can h be lower than P^d.

Finally, suppose that P_n^d is the value of P^d of sampling period n, the detection rule is given by

$$\delta_n = \begin{cases} 1 & if \quad P_n^d \geq h_n \\ 0 & if \quad P_n^d < h_n \end{cases} \tag{5.8}$$

where δ_n is also an indicator function of whether there is attack existing in the network during the sampling period.

5.1.3 Performance Evaluation

5.1.3.1 Simulation Setup

Like in Sect. 4.4, we set up a VoIP network through computer simulation. In the normal traffic condition, the SIP INVITE generating rate is uniformly distributed from 25 per second to 75 per second with a mean of 50. The senders of these messages are randomly chosen from 100 normal users. Using the normal traffic as background, we inject malicious traffic from an attacker who is able to manipulate its INVITE generating rate. For the sketch traffic summarization, we set the sampling period $\Delta t = 10$ s and the table size $K = 32$ to achieve high detection resolution and low computational cost. For the threshold estimation parameters, we set $\lambda = 2.5$ and $\mu = 1$ as they are sufficient to capture the deviations brought by attacks without generating false alarms.

5.1.3.2 Detection of Traditional SIP Flooding Attacks

We first check the capability of the wavelet-based scheme to detect the traditional flooding attacks, where the attack rate is suddenly brought up to a certain degree. We inject two attacks with the constant rate of 50 INVITEs per second, both of which

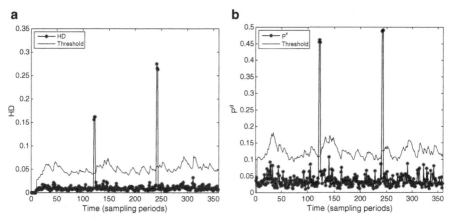

Fig. 5.1 Detection of traditional flooding. (**a**) Using HD; (**b**) using P^d

last for 30 s. Figure 5.1b shows the dynamic of P^d and the associated threshold. The two spikes clearly identify the two instances of attacks. Also due to the "threshold freezing scheme", the durations of the attacks, which are three sampling periods each, are precisely determined. Compared to Fig. 5.1a, we see that the wavelet-based scheme has comparable performance with the Hellinger distance (HD) based detection scheme from Chap. 4 when dealing with traditional flooding attacks.

5.1.3.3 Detection of Stealthy SIP Flooding Attacks

We inject one instance of the stealthy flooding attack where the attacker gradually increases its INVITE generating rate by 5 every 30 s, i.e., $\Delta r = 5$ and $T = 30$, from the initial rate $r_0 = 0$. The attack starts from the 180th sampling period and lasts for 300 s.

Figure 5.2a shows the result when we apply the HD-based detection scheme against the attack. We can see that the attack totally tricks the scheme. First, as it does not bring great changes to HD, the attack is able to prompt the threshold higher rather than driving HD to exceed the threshold. Second, when the attack ends, it causes sudden decrease of the overall traffic rate in the network and consequently results in great changes to the traffic distribution. Therefore HD becomes suddenly high and the threshold ironically keeps freezed after traffic becomes normal (see Sect. 4.3.2.2), leading to persistent false alarms. Thus the HD-based scheme is ineffective against the stealthy attack.

Figure 5.2b shows how the wavelet-based detection scheme responds to the stealthy attack. As illustrated in the figure, P^d is very sensitive to the attack and has a sudden sharp increase almost right after the attack starts. Also, P^d is able to come down immediately to normal values after the attack ends as energy corresponding

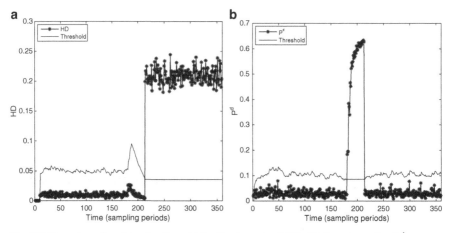

Fig. 5.2 Detection of stealthy flooding. (**a**) Ineffectiveness of *HD*; (**b**) detection using P^d

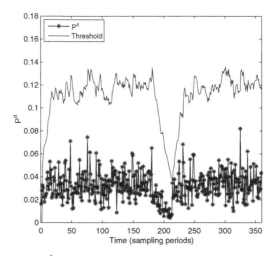

Fig. 5.3 Ineffectiveness of P^d on stealthy flooding without sketch

to the detail signal can not be affected by the attack any more. Due to the "threshold freezing scheme", we can again precisely determine the attack duration of 30 sampling periods, i.e., 300 s. Overall, the wavelet-based scheme preserves the ability to quickly and accurately detect the stealthy flooding attack.

5.1.3.4 Impact of Sketch on Detection

Sketch provides fixed-size manageable raw traffic signals to the wavelet-based attack detection regardless of the number of users in the network. Despite addressing the scalability issue, sketch also has crucial impact on detecting the attack. Figure 5.3 shows the dynamics of the detector P^d in responding to the same stealthy attack as described above without taking the raw signals from sketch. Apparently P^d loses its sharpness in detecting attacks when sketch is not in place.

5.2 Resource-Drained Malformed Message Attack Detection

5.2.1 Resource-Drained Malformed Message Attack

SIP provides open-ended control message implementation to allow including additional capabilities to enhance itself over time. However, one negative impact of such an open design is that it provides opportunities for attackers to understand the new implementation and take advantage of its vulnerabilities to initiate attacks. Realizing this, we identify a resource-drained malformed message attack.

The basic SIP as described above does not require the proxy servers to track states for already established sessions. The result is that the proxy servers are not able to determine whether a session is alive or dead. If a session fails without sending a BYE message, the proxy server will retain its reserved resource and have no way to release it. To resolve this issue, a SIP extension header field "Session-Expires" [7], or also called session timer is proposed as a keep-alive mechanism. When initiating a session, a UAC specifies its desired session duration through this timer included in the first INVITE message. Both the proxy server and UAS will agree with this timer if it is not too small through the handshake. The proxy server reserves resource for this session according to the timer. The resource will be released after the session timer expires or can be refreshed with a SIP UPDATE message. The UP-DATE is not discussed in this book as attackers disconnect from the network almost right after the session has been established and are unlikely to use the refreshing mechanism.

In normal situation, the session timer is utilized by SIP user agents to convey the durations of the sessions to the proxy server. But according to the standard of this SIP extension [7], only the minimum value for the timer is defined by the header "Min-SE". As a result, since no maximum value is enforced, an attacker can manipulate the session timer and specify its duration as long as he wants in the SIP INVITE messages to initiate the attack. The attacker also acts as both UAC and UAS himself, thus no third party would suspect these INVITEs. The proxy server will then reserve resource for the attacker according to the session timer. As the proxy server is not able to directly determine whether an established session is still alive or dead before the session timer expires, it allows the attacker to simply disconnect from the network but still hold resource reserved for the session on the proxy server. And wireless networks make the disconnection even easier. After the disconnection, the attacker will be silent for a while and then initiate another session with an arbitrarily long timer to further reserve resources on the proxy server.

Such an attack causes damages to the proxy server and gives the attacker advantages in two folds. First, to maximize the effect of the attack, the attacker will set the session timer to a large value and reserve resources as long as possible on the proxy server. Wireless networks allow the attacker to easily disconnect from the network and the disconnection does not release the resources until the session timer expires. Thus the attacker can repeatedly reserve new resource besides the resources already held by him each time he initiates a session. As the attacker continues initiating sessions and disconnecting from the wireless network, he is able to hold multiple resources simultaneously on the proxy server and prevent normal users from getting access to the service. Second, the attacker keeps silent for a while after each disconnection and initiates the attack following the normal traffic rate. Thus it can hardly be detected by existing volume-based intrusion detection systems. The attack virtually costs no extra resource of the attacker other than simply setting a value of the "Session-Expires" header each time before initiating a session. More severely, if several distributed attackers collaborate to initiate the attack, resources on the proxy server will be drained at a much faster pace. Overall speaking, the identified attack can easily consume a considerable amount of processing capacity of the SIP proxy servers, and significantly downgrade the servers' performance to process normal messages and manage sessions. To the best of our knowledge, this resource-drained DoS attack has not been addressed in the literature and we are the first to identify it.

5.2.2 Detection Scheme Design

5.2.2.1 Hypothesis Test and Observation Distributions

Our goal is to develop a scheme to efficiently detect the resource-drained attack. The main and most important characteristic of the attack behavior is unpredictable, which makes the detection problem very difficult. However, at a high level, the problem can be considered as making a decision on whether the attack exists or not in the network. Thus by observing the values of the session timers from the whole network, we design the detection scheme following the statistical hypothesis test approach.

Let $\{T_n, n = 0, \ldots, k\}$ be a sequence of the session timer values observed from the network. The observation point can be the proxy server as it is the target of the attack and also monitors every going-by SIP message. We consider two hypotheses, H_0 and H_1. The null hypothesis H_0 corresponds to no attack in the network whereas the alternate hypothesis H_1 corresponds to attacks existing in the network. Then the problem can be formulated as

$$Decide\ Between \begin{cases} H_0 : T_1, \ldots, T_k \sim f_0 \\ H_1 : T_1, \ldots, T_k \sim f_1 \end{cases} \tag{5.9}$$

where f_0 is the distribution of the session timer values when there is no attack whereas f_1 is the distribution when attacks exist in the network. We will choose between the two hypotheses based on the actual observed distribution of the session timer values.

The session timer conveys the durations of the sessions and there is no default value defined for the timer [7]. An efficient normal session timer setting should reflect the characteristic of the session holding time and thus avoid frequent refreshments to increase the signaling burden. Therefore f_0 can be modeled to have the same distribution of the session holding times, which is log-normal to reflect the long tail characteristic according to existing research [12]. However, since the attacker behavior is unpredictable, the distribution f_1 of session timer values under attack can hardly be characterized. Thus it is necessary to use a non-parametric approach where no assumption on f_1 is required to perform the detection, which will be discussed in the following.

5.2.2.2 The Anderson–Darling Test Based Detection Scheme

The Anderson–Darling (A–D) test [1,24] is a powerful statistical test tool which can examine whether the actual observed distribution differs from the null hypothesis distribution f_0. For the resource-drained attack detection, we observe the session timer sequence $\{T_n\}$ of size k and prepare it for the Anderson–Darling test.

We first sort the logarithms of $\{T_n\}$ from low to high and obtain another sequence $\{X_n\}$ of the same size. Next we will use the A–D test to check the normality of $\{X_n\}$. Let μ and σ be the mean and standard deviation of $\{X_n\}$ respectively. We then standardize $\{X_n\}$ to get $\{Y_n\}$

$$Y_n = \frac{X_n - \mu}{\sigma}. \tag{5.10}$$

This standardization allows us to perform the normality test without knowing the specific parameters for the normal distribution. Then with the cumulative distribution function (CDF) of the standard normal distribution Φ, the Anderson–Darling test statistic is

$$A^2 = -k - S \tag{5.11}$$

where

$$S = \sum_{n=1}^{k} \frac{(2n-1)}{k} [\ln \Phi(Y_n) + \ln(1 - \Phi(Y_{k+1-n}))]. \tag{5.12}$$

Then an approximate adjustment of A^2 for the sample size k is calculated to avoid skewness [24]

$$A^{*2} = A^2(1 + \frac{4}{k} - \frac{25}{k^2}). \tag{5.13}$$

Finally, we have the detection stopping rule

$$Choose \begin{cases} H_0 : A^{*2} \leq \beta \\ H_1 : A^{*2} > \beta \end{cases} \qquad (5.14)$$

where β is the critical value of the Anderson–Darling normality test and the choice of β's actual value is based on the significance level we want to achieve [24]. From (5.14), we see that the attack is detected when A^{*2} exceeds β. Algorithm 2 describes the detection scheme on a session timer sequence of size k.

Algorithm 2: A–D test based detection scheme with critical value β

---

1. Record k observations $\{T_n\}$ of the "Session-Expires" header values from the users in a network.
2. Calculate the logarithm of every element in $\{T_n\}$, sort them from low to high, and obtain the sequence $\{X_n\}$.
3. Standardize $\{X_n\}$ to obtain $\{Y_n\}$ using (5.10).
4. Calculate the Anderson-Darling test statistic A^2 from $\{Y_n\}$ using (5.11).
5. Adjust A^2 for the sample size k to obtain the adjusted test statistic A^{*2} using (5.13).
6.
if $A^{*2} > \beta$ **then**
    reject H_0. The attack is detected.
else
    do not reject H_0. No attack is detected.
end if

Another powerful widely-used statistical test is the Kolmogorov–Smirnov test [3]. However, we choose the A–D test over the K–S test based on two of its merits. First, the A–D test is more sensitive at the tail of the distribution. Second, the A–D test is especially powerful at test for normality when the parameters of the reference normal distribution are not known. Both of the merits tally with the characteristic of f_0, which justifies our choice of the A–D test.

5.2.3 Performance Evaluation

In this section, we evaluate the performance of the A–D test based detection scheme on the novel resource-drained attack through computer simulation. The performance is measured in terms of the number of samples to reach the detection decision under different attack conditions. Besides showing its ability to detect the basic one attacker attack, the scheme works even more effectively when dealing with the DDoS attack.

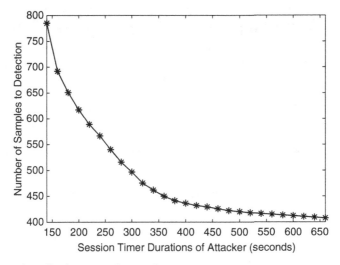

Fig. 5.4 Detection of basic one attacker attack

5.2.3.1 Simulation Setup

We consider a wireless VoIP network with 60 users, who can be either attackers or normal users. The normal traffic is simulated based on the data set from [12]. Each user sets a value for the "Session-Expires" header field before sending out the INVITE message to initiate a SIP session. In the normal traffic, the session timer values follow a log-normal distribution, where the mean is 116.75 s and the standard deviation is 262.28 s. The critical value β is set to 0.751 to achieve a significance level of 5% [24]. The detection is performed according to Algorithm 2 each time a new sample is picked in. The normal data set passes the test even when the sample sizes are small thanks to the approximate adjustment of the test statistic and the significance level of 5%.

5.2.3.2 Detection of Basic One Attacker Attack

In this attack case, there is one attacker who can manipulate his own session timer values as long as he wants among the total 60 users. As the mean of the normal session timers is about 120 s, the minimum timer of the attack is set to 140 s for the attacker to gain some advantage over normal users. The severity of the attack goes up with the values of the attacker session timers. The traffic arrival of the attack is in conformance with the normal traffic, thus obviously it can not be detected by volume-based intrusion detection systems. We monitor the values of the session timers from all the users. Even though some INVITE messages sent by the users do not result in established sessions, we still include their session timer values into the test since they indicate the attempts of initiating sessions. Figure 5.4 shows the

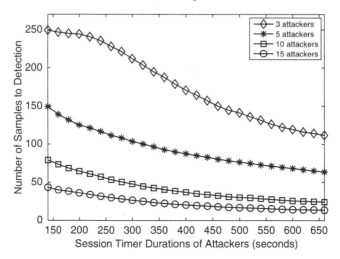

Fig. 5.5 Detection of DDoS attack

average detection delay in terms of the number of samples to reach the detection decision. The clear trend from the figure is that the longer the attacker wants to hold the resources on the proxy server each time, the faster it will take to detect the attack. Take the attack of 400 s session timer for example, about 436 samples from the whole network are needed to detect the attack. However, considering the attacker sends his attempts at the frequency of about every 60 INVITE messages of the whole network, the attack is detected around the 7th attempt of the attacker, where no severe damage has been done to the proxy server yet considering the processing capability of even open source SIP proxies [22].

5.2.3.3 Detection of DDoS Attack

To increase the intensity of the attack and hold resources on the proxy server as fast as possible, one option for an attacker is to increase his attacking rate. Nevertheless, this approach also increases the traffic volume of the attacker and is very likely to be detected by volume-based intrusion detection systems. The main damage of the resource-drained attack identified in this work does not depend on the sending rates of the attack attempts, but on the duration of the manipulated session timer in each attempt. Instead of increasing one attacker's attacking rate, multiple distributed attackers can collaborate to initiate the attacks together. The attackers all set long session timers themselves and collaboratively initiate attacks to drain the resources on the proxy server at a much faster pace. Note that in this case the traffic arrival of all the attackers still follows the normal traffic arrival but collectively they can significantly increase the intensity of the attack and still escape the volume-based detection. Figure 5.5 shows the average detection delay of the DDoS attacks. Attack cases of 3, 5, 10 and 15 attackers out of the total 60 users are considered in our simulation. Each attacker sends the same form of the resource-drained attack and collaboratively they can hold resources on the proxy server multiple times as fast as the one attacker case based on the number of attackers in the attack. From Fig. 5.5, besides the trend that longer attacking session timer causes quicker detection, we see that the DDoS attacks are detected much more quickly than the one attacker attack of the same session timer durations. Also, the more the attackers there are, the quickly the attacks will be detected. Thus the detection scheme is more effective when dealing with the DDoS attack.

5.3 Summary

In this chapter, we first develop a wavelet-based detection scheme to address the stealthy SIP flooding attack. Wavelet allows us to extract information from the raw traffic signal by decomposing the signal into different levels. Thus the detector identified by us, namely the percentage of energy corresponding to the detail signal, is able to quickly expose the changes brought by the stealthy attack even though the attack only slowly influences the traffic. Also, we utilize the sketch technique to provide fixed-size manageable raw traffic signals as the input to the wavelet-based detection regardless of how many users exist in the network. Through computer simulation, we demonstrate the ability of the wavelet-based scheme to quickly and accurately detect the stealthy attack.

Then, we identify a novel resource-drained malformed message attack which works by exploiting vulnerabilities of one SIP protocol extension and wireless networks. The "Session-Expires" header, or the session timer, is originally proposed as a keep-alive mechanism for SIP. However, it provides attackers opportunities to reserve resources on the SIP proxy servers as long as they want. Also, wireless networks allow attackers to easily disconnect from the network and the disconnection does not release the resources on the proxy servers until the session timer expires. Attackers can hold multiple resources through repeating reservations and disconnections, and greater damages can be caused if collaborative attacks are initiated from distributed attackers. As a counter-measure, we develop a robust detection scheme for the resource-drained attack based on the statistical Anderson–Darling test through investigating the characteristics of both the normal and attack behaviors. The scheme utilizes the changed statistical property of the session timers induced by the attack as the key insight that leads to detection. Through computer simulation, we show that besides the capability to detect the basic one attacker resource-drained attack, the scheme is even more effective when dealing with the DDoS attack.

5.4 Related Work of SIP Layer Attack Detection

VoIP is one of the most prominent and representative applications of IP-based multimedia communications. Surveys of the VoIP security issues can be found in [9, 10, 23, 25, 26]. SIP [19] is one fundamental component of IP-based multimedia communications, and the SIP flooding attack is among the most severe of all because it is easy to launch and capable of quickly draining the resources of both network and nodes. To detect the flooding attack, a natural idea is to identify changes in traffic volume or rate [14, 15, 20]. In such schemes, alarms are raised if the traffic volume during a time interval is larger than a threshold predicted according to past normal conditions. A main issue of volume/rate monitoring is that the detection accuracy can be severely degraded if the normal rate is dynamic in the observation window (due to the random nature) and the flooding attack rate is not very high. Considering that the low-rate flooding is likely to introduce different probability distributions from the normal traffic, several statistical-based schemes have been proposed to address the problem, e.g., the Hellinger distance [21, 27] and entropy [16, 17]. In this book, we utilize the Hellinger distance integrated with sketch [5, 11] (a data aggregation technique) for low-rate SIP flooding detection. The proposed scheme can accurately detect low-rate attack and maintain a dynamic threshold to reflect normal traffic behavior. Further realizing that a stealthy attacker can slowly pollute the system and evade detection, we also study the stochastic model of stealthy attackers and develop efficient detection methods correspondingly.

The schemes presented in [4, 8] also work effectively to detect the SIP flooding attacks. In their work, SIP transactional models are built to detect deviations from normal behaviors. However, these schemes are customized specifically to the SIP protocol suite and can not be easily generalized to other flooding detection cases. Whereas in our scheme, we can use the attributes associated with protocols other than SIP to profile traffic and thus have a generic method to detect flooding attacks.

Wavelet analysis has been used in [2, 13, 18] for network anomaly or DoS attack detection, which utilize the coefficients transformed from the original traffic signal to identify changes induced by the attack. However, none of them explore wavelet's power to detect the stealthy flooding attack.

The malformed message attack is another major problem in SIP-based VoIP networks [25]. SIP defines many open-ended control message implementations to allow including additional capabilities over time. The "Session-Expires" extension header is proposed in RFC 4028 [7] as a keep-alive mechanism for SIP. In the "security considerations" section of the RFC, only attacks through setting very small session timers are addressed, where an attacker may force compliant user agents to frequently send session refreshes at a rapid rate. The RFC proposed a 422 (Session Interval Too Small) response to reject the attacker's request if the timer is smaller than the value specified in the "Min-SE" header. However, there is no enforcement of how large the session timer can be and we identify a novel resource-drained attack by utilizing this fact. Also, the victim of our identified attack is the SIP proxy server rather than the user agents as considered in the RFC. Before our study, such kind of malformed message attack has not been addressed in the literature.

References

1. T. Anderson and D. Darling, "Asymptotic Theory of Certain "Goodness-of-Fit" Criteria Based on Stochastic Processes," *Annals of Mathematical Statistics*, 1952.
2. G. Carl, R. Brooks and S. Rai, "Wavelet Based Denial-of-Service detection," *Computers & Security*, vol. 25, no. 8, pp. 600–615, Nov. 2006.
3. I. M. Chakravarti, R. G. Laha, and J. Roy, *Handbook of Methods of Applied Statistics, Volume I*, John Wiley and Sons, pp. 392–394, 1967.
4. E. Chen, "Detecting DoS Attacks on SIP Systems," in *Proc. 1st IEEE Workshop on VoIP Management and Security*, 2006, pp. 53–58.
5. G. Cormode and S. Muthukrishnan, "An Improved Data Stream Summary: The Count-Min Sketch and its Applications," *J. Algorithms*, 2004.
6. I. Daubechies, *Ten Lectures on Wavelets*, Philadelphia, PA: SIAM, 1992.
7. S. Donovan, and J. Rosenberg, "Session Timers in the Session Initiation Protocol (SIP)," IETF RFC 4028, Apr. 2005.
8. S. Elhert, C. Wang, T. Magedanz and D. Sisalem, "Specification-Based Denial-of-Service Detection for SIP Voice-over-IP Networks," in *Proc. 3rd IEEE International Conference on Internet Monitoring and Protection*, 2008, pp. 59–66.
9. D. Geneiatakis, G. Kambourakis, T. Dagiuklas, C. Lambrinoudakis and S. Gritzalis, "SIP Security Mechanism: A State-of-the-Art Review," in *Proc. 5th International Network Conference*, 2005, pp. 147–155.
10. D. Geneiatakis, T. Dagiuklas, G. Kambourakis, C. Lambrinoudakis, S. Gritzalis, K. S. Ehlert and D. Sisalem, "Survey of Security Vulnerabiliteis in Session Initiation Protocol," *IEEE Communication Surveys & Tutorials*, vol. 8, no. 3, pp. 68–81, 2006.
11. A. Gilbert, S. Guha, P. Indyk, S. Muthukrishnan and M. Strauss, "Quicksand: Quick Summary and Analysis of Network Data," DIMACS Technical Report 2001–43, 2001.
12. F. Gustafson and M. Lindahl, "Evaluation of statistical distributions for VoIP traffic modelling," University Essay from University West, Department of Economics and IT, 2009.
13. C. Huang, S. Thareja and Y. Shin, "Wavelet-Based Real Time Detection of Network Traffic Anomalies," in *Proc. Securecomm and Workshops*, 2006.
14. B. Krishnamurthy, S. Sen, Y. Zhang and Y. Chen, "Sketch-based Change Detection: Methods, Evaluation, and Applications," in *Proc. ACM SIGCOMM IMS*, 2003.
15. A. Lakhina, M. Crovella and C. Diot, "Diagnosing Network-Wide Traffic Anomalies," in *Proc. ACM SIGCOMM*, 2004.
16. A. Lakhina, M. Crovella and C. Diot, "Mining Anomalies Using Traffic Feature Distribution," in *Proc. ACM SIGCOMM*, 2005.
17. X. Li, F. Bian, M. Crovella and C. Diot, "Detection and Identification of Network Anomalies Using Sketch Subspaces," in *Proc. ACM IMS*, 2006.
18. W. Lu, M. Tavallaee and A. Ghorbani, "Detecting Network Anomalies Using Different Wavelet Basis Functions," in *Proc. Communication Networks and Services Research Conference*, 2008.
19. J. Rosenberg, H. Schulzrinne and G. Camarillo, "SIP: Session Initiation Protocol," IETF RFC 3261, Jun. 2002.
20. R. Schweller, Z. Li, Y. Chen, Y. Gao, A. Gupta, Y. Zhang, P. Dinda, M. Kao and G. Memik "Reverse Hashing for High-Speed Network Monitoring: Algorithms, Evaluation, and Applications" in *Proc. IEEE INFOCOM*, 2006.
21. H. Sengar, H. Wang, D. Wijesekera and S. Jajodia, "Detecting VoIP Floods Using the Hellinger Distance," *IEEE Trans. Parallel Distrib. Syst.*, vol. 19, no. 6, pp. 794–805, Jun. 2008.
22. SIP Express Router, [Online.] Available: `http://www.iptel.org/ser/`.
23. D. Sisalem, J. Kuthan and S. Ehlert, "Denial of Service Attacks Targeting a SIP VoIP Infrastructure: Attack Scenarios and Prevention Mechanisms," *IEEE Network*, vol. 20, no. 5, pp. 26–31, 2006.

24. M. Stephens, "EDF Statistics for Goodness of Fit and Some Comparisons," *Journal of the American Statistical Association*, vol. 69, pp. 730–737, 1974.
25. VoIPSA, "VoIP Security and Privacy Threat Taxonomy," Public Release 1.0, 2005.
26. S. Vuong and Y. Bai, "A Survey of VoIP Intrusion and Intrusion Detection System," in *Proc. IEEE 6th International Conference on Advanced Communication Technology*, 2004, pp. 317–322.
27. G. Yang and L. Le Cam, *Asymptotics in Statistics: Some Basic Concepts*, second edition, Wiley, Mar. 2006.